AF366805

Series

Accademia

Tiziano Cantalupi

Filippo Onoranti

THE BIRTH OF THE UNIVERSE FROM NOTHING

Series **Accademia**
The birth of the Universe from Nothing
by Tiziano Cantalupi and Filippo Onoranti
First English edition: 2024
© 2024, Santelli editore

Gruppo Editoriale Santelli
Santelli editore *dal 1987*

www.santellieditore.it
www.grupposantelli.it
Via F. Filzi, 3
Cinisello B. - Milano - 20092
340 948 1047

Tiziano Cantalupi: tcantalupi@libero.it
Filippo Onoranti: filippo.onoranti@gmail.com

PREFACE

Dear reader, what we have tried to outline in this work is a guideline for a journey that we have all undertaken living among Nature. Together we wanted to share our outlooks, intuitions and considerations on the elements at the edge of this dimension, in which our existence unfolds; we did it by taking into consideration two observation points, which apparently seem very distant, but are close in taking seriously all the impossible questions that appeared along the way.

In the first part of the book, we addressed issues related to the conception of the Universe, the concepts of Being and Nothingness, going in search of the first traces that humanity left of this debate and that still to this day – we can anticipate this – we know underlie our way of speaking, reflecting and looking at the world around us. We will also provide comparisons which have been milestones of the journey in search of the origins of Nature by committing ourselves to underline that in addition to the metaphors and tools that may sometimes appear outdated, man's gaze towards the big questions has not changed much since we are able to leave a trace for posterity of our questioning. We like to think that this common destiny makes us all a little closer and represents a foundation of our human being. The awareness of being will be one of the fundamental elements with which we will deal and will represent a decisive theme also to clarify the possible meanings of Being and Nothingness.

Having chosen to fix our gaze on some of the most radical themes that it is possible to embrace with the mind, we even haven't left behind the matter of God, which has a lot to do with Being, with Nothingness and with the birth of the Universe. The guideline we gave ourselves to share our ideas on the subject was to read it on a level that attempted to remain as free as possible from fideistic or cultural interpretations, also attempting to look at the God of philosophers, if you will, ancient and modern, rather than at the God of traditions. The reasons for this choice were mainly two: the desire to make the conversation as open and inclusive as possible, without binding it to particular theological traditions; and the conviction that the question around the absolute, the search for what tries to find an answer in the name of God, is a profound and indispensable need of the human soul and which therefore goes beyond the distinction between science and faith, and even more between faith and faith.

Before leaving room for the more technical and rigorous discussion on a "scientific" level, we wanted to offer a look at the possible ways of interpreting the relationships between Nothingness, Being and God, on a logical level and in a broad philosophical sense.

In the second part of the book, starting from an unusual hypothesis, we presented a picture of a completely self-sufficient universe; a universe that can generate itself starting from nothing and from the laws of physics. It must be said, for the sake of truth, that in the scientific landscape there are already several publications (books and articles) that deal with the birth of the universe starting from the "quantum vacuum". Some publications have a remarkable intrinsic scientific substance, others, however, despite having had significant diffusion, base their theses on a clear misunderstanding.

In all the publications just mentioned, however, the mechanisms that account for the existence of the laws responsible for the universe's appearance are not sufficiently explored. The existence of regulating laws is almost taken for granted without specifying their "origin". Furthermore, the laws themselves are relegated to the role of supporting actors.

The role of laws, from this work's point of view, however, is fundamental since once they "emerge" from nothing, any "entity" (universe or other entity equivalent to it) can take shape. Laws are the real engine for the appearance of "something", they precede the appearance of space, time, energy and matter itself. The laws that require our universe to come into existence, develop and evolve; we, ourselves, are also an inevitable by-product of such laws.

In the Second Part of the volume, therefore, we tried (also by resorting to original formal hypotheses) to give a satisfactory explanation of the mechanism of the appearance of these laws. In particular, we start from the idea that the laws, or mathematical relationships, are simply potentially "suspended over nothing" and that, transcending physical reality, they have infinite possibilities of expressing themselves; to pass from potency to act. If we start from the idea that there are potentially infinite possibilities that a given "order" can randomly or spontaneously arise from nothing, nothing prevents that something like a law can manifest itself.

If the laws are a potentially infinite number, our universe can only be one of many potentially possible universes. A universe like ours could have originated according to the laws we know; other universes could have arisen following opposite laws and this completely by chance. The same fundamental constants of our universe are such because the laws of physics that allowed the universe to be born have randomly taken the "shape" that we know.

We can therefore say that in addition to ours there are other universes, in fact there are potentially infinite universes, and we are on one which allowed our life, because it is among those whose laws include the quantum theory. The truth is that the fact of existing does not make a particular universe special since in the multiverse all possible universes find their place. To return now to what has been said regarding the publications produced so far on the universe born from the "quantum vacuum", it is possible to add that in these works it is hypothesized that the mechanism of appearance of a universe from the vacuum takes place (by exploiting the Uncertainty Principle) by quantum fluctuation. At the basis of these works, however, there is a subtle misunderstanding concerning the concept of emptiness and nothingness; in fact, it is not specified that the void is something other than nothing. The quantum vacuum, in fact, unlike nothing, is teeming with fields, virtual particles and various fluctuations. Therefore, the hypothesis of a universe born from the quantum vacuum should be preceded by the explanation of how the vacuum was formed starting from "nothing".

The aspect concerning the importance of defining the exact conditions in which our universe originates is overcome in this volume by resorting to the possibilities that quantum formalism offers regarding a particular phenomenon called "superposition of states". In quantum mechanics the state of emptiness (nothingness) can also be thought of as a superposition of states that are not empty. Practically, the void, the true nothingness, can be represented, according to the interpretation given in this book, as a superposition of a "void" and a "non-void". It is not easy to form a mental image, above all if we think within the context of the dual logic underlying daily life, of an "entity" which is at the same time something and its opposite, but quantum mechanics allows for these bizarre logics in the folds of its formalism. So, from the overlapping of "void" and "non-void" it is possible, indeed, it is certain, that a "non-void" or something can "spontaneously" emerge. Something – for example a field – which can be the seed for the birth of a universe like ours.

To the well-known question "why is there something instead of nothing?", one can therefore answer that "something always inevitably arises out of nothing".

During the drafting of the Second Part of this book, where a completely self-sufficient universe was presented, a truly curious aspect emerged regarding the role of God. In fact, God could have chosen that the universe would be born from nothing, that it would be generated from itself. With God's exit following the outburst of science, he would re-enter through the window following a deliberate divine choice.

At the end of the Preface, a warning for the reader. The exposition of certain concepts concerning quantum theory has not allowed to avoid certain formal passages, the reader can however avoid dwelling on these points without losing what is the global meaning emerging from the individual sections of this volume.

PART ONE

NOTHINGNESS, GOD, UNIVERSE

Filippo Onoranti

Chapter I

WHICH QUESTIONS ARE WORTH ASKING?

1. Introduction

Where do we come from? This question has tormented every culture that has ever existed on our planet; maybe, in some other corner of the universe someone, in this exact moment is asking himself the same question. If you have chosen to follow us on this journey, probably you have asked yourself more than once the same. To us it seems a wonderful query, on one hand because it made us come to each other, on the other because, as many other unsolved riddles, it leaves each one of us with that small precious freedom to interpret the clouds above the world by giving them any shape we like and that reassure us.

Since the responsibility of answering this question is impossible, we decided to spend time with those who devoted their lives to this mission – literally – on the brink of reality. Artists, scientists, philosophers and mystics, in front of these matters we are all in the same boat, but we like to think that whoever has taken us until here, maybe had something we could learn from, some suggestions to make the path more comforting; or that could suggest us a glimpse that they may have seen by chance, but from which they have been swallowed. These first pages are in fact dedicated to the stories of these travellers between Being and Nothingness.

We want to consider tradition, that brought us here as if it were man's unconsciousness, like a family story, in which the great are friends who came before us, and philosophy is nothing more than a free attempt in thought's history.

Another main subject, mainly covered in the second part of this book, will be Science, which will tell us the same story but from another point of view. In any way, the goal is to perceive the world as less distant, to feel more like its friend rather than a guest. This second point of view, that to simplify we can call "scientific", is knowledge, and it's an intellectual act though which we relate an experience to a criterion, and that has a lot to do with measurement and evidence. Therefore, knowledge looks towards the horizon of phenomenon, and it involves the possibility to confront ourselves to debunk, or not, our beliefs. On the other hand, it's oriented towards something that transcends phenomenal reality, and that the philosophical tradition has sometimes called Being and sometimes God, often striving to build bridges between these two terms; the villain of this story is played instead by the Nothingness, which escapes understanding as much as experience, but in addition makes us feel alone and as if the ground were missing from under our feet. Who knows if it might not for this very reason teach us to need it less and stimulate us to open the wings of our mind.

It is a necessary clarification to make before starting the journey which issues are part of this second dimension and which are part of the first. Understanding the nature of one's purpose, argumentative, intellectual, or cognitive is necessary to move steady steps through this discourse. The "rules" of metaphysics are in fact not the same as those of physical (or phenomenal) reasoning and knowing which language to speak is necessary to understand each other, even if only with oneself.

Being and Nothingness are landscapes of thought; therefore, it is not on the path towards knowledge that we will encounter them. Us Human Beings, simply as entities, find ourselves on an immeasurable level – literally impossible to relate – to those dimensions that constitutively transcend us. We move within experiences, and we are phenomena; Being and Nothingness are, respectively, totality and absence of the same possibility that phenomena and experience exist. Among these phenomena there are also our thoughts, and the language which gives them shape and which makes us comprehensible. It seems like a pun, but it's not. Trying to think about a feeling, which exist regardless of its name, but how difficult would it be to distinguish tears of sadness from tears of happiness? If we were only sensations the difference could slip away. We often need memories and experiences to discover how to distinguish the nuances of our thoughts and feelings. Because if emotions are closer to our flesh, feelings are complex phenomena instead, halfway between the world of the mind and the world of the body (if this distinction makes any sense, but this is another journey to discover). Nothingness, perhaps more than Being, offers itself as a feeling aroused by the idea of absence to which imagination or reason add the nuance of absolute, but surely not as something that can be experienced.

Thus, languages think in us, and offer us tools to measure, unite and cultivate our thoughts, that without words would often be unintelligible. Words that often we haven't chosen, but that have been given to us to become pathways, sometimes walls other times bridges, in order to create our map of the world, both the material one and the many immaterial worlds in which we live. So – anticipating as if we were looking at a photograph of exotic destinations, what we will be exploring in a few pages – Being and Nothingness are moods perhaps even before designating things or concepts, they are names we give to our finding ourselves in front of what we feel are our roots or what is completely unknown. Being and Nothingness respond in some way to the desire to move from the level of knowledge and from coming across entities towards thought; a desire that Heidegger – who will sometimes be our companion – bases on the freedom of turning from things to their origin, attempting to see what makes them possible beyond what they are or are not.

We don't know what the first step of our journey is; probably long before written languages made their appearance on the world, around their hearths, our ancestors had been handing down stories about the origins of their lineages for millennia and trying to give an understandable form to that nature in which they were merged. These stories have not reached us; at least not in their ancient form, but the impetus that animated them is still alive in that feeling that we can call spirituality and which in Greece, where this adventure of ours will begin, took the name of philosophy. But first we want to share an anecdote, one of those that have the flavor of legend, to introduce us to the climate that will accompany us from here on, and which precisely concerns the origins of civilization.

By asking this question to experts and scholars of the most disparate disciplines, you will certainly get just as many answers, but the one that struck us the most is offered by the American anthropologist Margaret Mead. Civilization begins with the first discovery of a healed fracture of a femur. This is because in nature, with such a wound, one cannot take care of oneself. Finding food or water, sheltering from severe weather and predators is impossible for weeks or even months, if not for the entire duration of one's life. This fossil scar testifies that someone took care of that wounded companion, fed him and welcomed him into their shelter. In short, it testifies to an act of deliberate humanity, the ability to take charge of the needs of others without a convenient reason, by the simple fact of recognizing a similar person. Today we could call it empathy, but whatever name we want to give to this feeling, it represents the basis of community life and is based on the ability of the human mind to go beyond mere selfishness, instinctive calculation of one's own profit and to carry out actions that constitute the opening of man onto another level opposed to the material one. That battered femur is the relic with which in the mists of time we opened the doors of the world to our human soul.

From that moment on, the narrative and the history of the world were entrusted, first to words, then to images and finally to writings in mythical form, in a way therefore aimed at affecting listener's sensitivity rather than their reasoning, to arouse emotions and to imprint them in the memory. The great ancient civilizations were built on these foundations. But in a small and harsh land in the lap of the Mediterranean, for the first time, something changed. The myth, with the power of its images and the liveliness of its symbols, was accompanied by a new way of telling, and therefore also of thinking about nature: philosophy was born. That was the moment in which our Western culture was also born. Man wanted something more than "simply" inhabiting nature, he wanted to try to understand it, to embrace it with his own mind even in those aspects that escaped his understanding. For this reason, the birth of philosophy is also the birth of science, and for centuries, as if they were sisters, they shared their childhood holding hands and not distinguishing themselves from each other. Even today their union, which became rarer with time, could help our modern civilization which began to take its first steps precisely from this union.

The goal was as clear as the sun and as ambitious as Icarus: the search for the truth. Aletheia, so called by the Greeks, is that which rests under a veil of oblivion or opacity. What is true is what lies behind appearances, behind mutability, which can account for it. The truth is what can save us from the abyss of Nothingness, from uncertainty. A desire that almost becomes a challenge for a people, also due to this more unique than rare aspect, devoid of any consolatory narrative on the theme of death. In fact, the Greek looks at the future as a synonym for decay and corruption; he perceives it as the emergence of Nothingness among beings, but instead of opposing it, he mostly accepts it, without welcoming it or celebrating it, simply taking note of it. Hades exists, and it is a place of pain and forgetfulness which does not correspond to celestial meadows or otherworldly dimensions.

The Greek looks at Nothing without looking away and without any filter, he takes note of it and takes charge of it in all its profoundness. This is how one of the protagonists of our journey entered the scene. It is not, as we have already recalled, an object of nature, or a phenomenon among others, but a metaphysical thought; a way of being of reality when this is completely alien to the relationship with what we are and feel. Nonetheless, on the threshold of this abyss, thought can take a step beyond the senses and at least try to get closer to the impossible.

Things and phenomena begin to prove insufficient. Although they are all that exists, this existence is not enough for us, since the human mind can challenge it; it cannot lead our steps beyond the threshold of phenomena, but we are given – if we do not limit ourselves to the senses of the body – to peek a little further. Thus, in Ancient Greece the search for the primary foundation of reality, the *arché*, principle of all things, after some initial identification with the natural elements, was sought beyond custom and became *Nous* – that is, divine thought – in Anaxagoras, and Apeiron, the indeterminate, the unlimited, in Anaximander, something in short that has nothing in common with the realities we know in ordinary experience.

Thought proves to be the most satisfying tool in this search.

Along the same direction, another crucial stage is marked by Heraclitus, who marks a definitive transition from phenomenal reality to that dimension which, centuries later, will be called metaphysics. The philosopher shifts attention from things to the reason by virtue of which they are what they are and nothing else; from the incessant changes that characterize every natural object, he suggests, and almost warns us, to shift our attention to what precedes them and welcomes them. Thus, the principle becomes *logos*. A word that has its origins in the land and in the rural daily life of the Greek people. In fact, it comes from the verb *legein*, a term used to indicate the action of gathering crops of the same type; it is what holds things together with a criterion, with a rule and not in bulk. And it is the true reason, the profound reason that lies behind the mere appearance, that holds reality together. It is its law before being its effects. And for this reason, it also indicates the word, because from this human thoughts and communities were born.

The first task entrusted to Adam was that of giving a name to things, so as to be able to recognize and grasp them with the mind even before with the body; and in an era almost contemporary with the Greek one that is hosting us, on the other side of the Eurasian continent, another witness to human history, Confucius, questioned on what his first action would be if he were entrusted with the task of governing the cities he replies that he would correct the use of names.

The word and the reason that this makes possible and communicable are recognized by men as an antidote to Nothingness, as a deadly and future instrument, yet effective and more resistant to the becoming of its bearers, to deal with transience. In the nothingness of death, the word manages to survive at least in part, making the mind and thoughts of its authors present, by having other men lend their voices in the narration, and then, in the encounter with the writings.

The nothingness of the unknown, the impossibility of embracing phenomena right down to their source with what comes from the senses, can be minimally filled by reason, which makes the possible an act, at least in thought. Thus a parent can dream of their children's adulthood, even if it is not certain that they will be able to be close to them with a hug, but they will be able to be part of their future in the outdated but possible form of memory; the scientist can push himself and see what will open up knowledge that is still immature but constantly growing, and the awareness that every answer opens up at least two questions makes him a friend of anyone who will continue the quest for knowledge. Each of us can, in the inexhaustible attempt to shed light on the unknown, recognize ourselves as part of the human community. Thus nothingness, what is immeasurable today, will move tomorrow, like waves on the beach that modify the profile of the coast, and sometimes discover unsuspected gifts.

2. The first traces of Being and Nothingness

As soon as we dock in the ports of Greece, Asia Minor or the Hellenic colonies in the south of Italy, we must be careful not to mention Nothingness – this path we have embarked – too lightly. In fact, this thought seems outrageous to the Greek; a contradiction in itself and even a provocation compared to the world's vision ancient men used to face nature.

Like many uncomfortable thoughts, our Nothingness remained taboo for several centuries, until, in Elea (near present-day Salerno), Parmenides tackled the matter. In his investigation of nature, he detached himself from his predecessors, as sometimes happens to those who experience the culture of their homeland from afar. He didn't look at the world with the senses, but with the mind alone, in an attempt to perceive something that was not at the mercy of change. This is how the science of being was born: ontology. Being which is defined with a lapidary sentence – and which perhaps resurfaces from high school memories like a tongue-twister – "Being is and cannot not be". Thus, Parmenides traces a furrow that will be a guide for the entire history of subsequent thought. But the counterpart to the Parmenidean aphorism sees Nothingness itself as the protagonist, or at least one of its names, as we will observe more closely in the following pages. In fact, Parmenides continues: "non-being is not and cannot be", doubly linking Being and Nothingness, as if one were day and the other night; but a day that never sets and a night that never enters the scene of reality. The former is the way of truth, the latter of error and deception. Man's enemy is doubt, and the mutability that feeds it, making the world that welcomes us elusive and changing, becomes something we try to deny and oppose.

Well before our Hellenic hosts, man was not so different, or at least not on this specific subject. They too, visiting the relics of ancient worlds as we do, were enchanted by their ancestors' attempts to oppose nothingness, death and the unknown. The checkmate that holds man has in fact always been the one between the finitude he perceives in himself and in his works, and the eternity he feels comes from nature – thought of as that immutable background that no god and no man made. Then temples and pyramids challenge and admire the unshakable nature of the mountains; art and beauty dare beyond the boundaries of corporeality alone to venture into the world of the spirit.

The same, but in other ways, is attempted by religions that seek coherence and meaning in the world, or even just symbols to make the distance between our mortality and the infinity of the cosmos less alien, less radical. This is precisely the path on which the sciences and philosophy will be born, less and less compliant with the mere transposition of narratives from the past and animated by the desire of each man to be a protagonist in the first person in the history of worldviews; to write at least a footnote to the book that man writes about nature and not just read it.

Still facing the Aegean Sea, a contemporary of Parmenides such as Heraclitus feels the same challenge, and while on the one hand he makes becoming and conflict the motor and almost the source of all things, on the other, he denies that becoming is the ultimate foundation of reality. He imagines it to be subject to a rule, a law, on which the unity of opposites is founded. Behind the becoming and therefore also the transience of the real is hidden, to the eye of those who have awoken their minds from the sleep of appearances, a stable principle: the *logos*. On one point at least, we feel we fully share Heraclitus' thought: war, at least the war against nothingness, is indeed a source from which humanity has drawn much of its energy and inspiration. Were it not that this Nothingness is the child of human reason itself, which, being able to conceive what is not there, and thus generating it in its own mind, offers itself and its own flesh as fertile ground for the Nothingness to become something.

Leafing through the pages of Parmenides' poem reveals another fragment that, like a treasure chest, holds and reveals much about its author. Handing it over to you, I imagine him on a summer evening strolling on the beach of his native Sicily, perhaps after a banquet to catch his breath from the festivities, or perhaps at the end of any day, under that ancient sky studded with lights that the glow of our cities has been hiding for almost two centuries now. That sense of infinity makes the philosopher's heart flutter as much as anyone else's and arouses that mix of awe and wonder that blurs the boundaries between happiness and fear. Like any of us, Parmenides dreams of a happiness that does not fall prey to the unexpected; he longs for something he can trust and rely on, and not finding it, he is seized by a thrill. Then he creates it. The Nothingness that upsets him, that non-being that cannot be and that is the path of error and deception has an antidote: "the heart that does not tremble at the well-rounded truth".

Parmenides thus discovers himself to be a man of our time, in search of that permanent centre of gravity that has been coming out of the radios since the 1980s, but which has been playing and making every man's heart tremble since before we were *Sapiens*. For if Being cannot be said and escapes language, so Nothing cannot be silenced. It is already too frightening to cloak it even in silence, and at the same time, men make it, in various ways, the protagonist and antagonist of their own stories and worldviews since they are able to express them.

If childhood ends when, still children, we begin to realise that our parents are people like us, and adolescence is therefore a period of turmoil and contention with respect to authority, so philosophy, born in the peace of the ordered cosmos and in search of its source, faces its own age of passage when it discovers Nothingness, and fearing it, denies it but also confronts it, and attempting to turn around to scrutinise it from all angles, weaves the history of our world, which, so far at least, seems not yet to have fully emerged from its adolescence.

Then the almost 27 centuries between Heraclitus and Parmenides and Martin Heidegger are not such a gulf. What builds a bridge is the unsurpassed desire of the human soul to make its way between Being and Nothingness. That of Heidegger, who writes at the end of the *Age of Technology,* even though almost all his works are composed by hand and by candlelight in his mountain retreat in the Black Forest, is a warning to the man of all future times not to delude himself that he can find the answer to this question of Being and Nothingness with the tools with which he searches for and makes his way through the things of the everyday world. Being and Nothingness are not *thrown,* as we human beings are, between entities but are a beyond entity; and it is given to us to open ourselves to either side, but on the condition that we do not embark on this journey with only our bodily senses. Sometimes by getting entangled on one side and forgetting the other; sometimes by resisting the flow as if our strength were an instrument compatible with the flow of the river of Being, we are swept away. Heidegger reminds us that the compass on this journey is not built with hands and things, but that we already keep it, originally, in our hearts and minds.

So how can we understand, confront, and perhaps even comfort each other in the face of these abysses? Humanity's journey seems to be along a narrow mountain path: on the one hand, the inaccessible peak of Being, whose summit, which escapes view, arouses tales and myths, becomes in tradition the home of the gods, the landing place of the *arché,* the gateway to other worlds; on the other, the abyss of Nothingness, where sooner or later everyone feels destined to fall, and, although they have never felt the thud of the fall, the fear is inseparable from this thought. And we, halfway between realms that elude us, cannot stop wondering about their nature, and try to make them less inaccessible, if not to our feet – and this is what science and technology do by building grappling hooks of various kinds, to offer us at least a slightly less distant view – at least to our minds. What is reconfirmed with each new endeavour to bring clarity is the impossibility of separating the two elements. It was at the height of industrial revolution that a metaphysician like Schopenhauer, to offer his own view of Parmenides' thought, wrote that we think of Nothingness only as the negation of Being, as *nihil* privative. Almost as if these two elements were an indeterminate pair, and that their product was what we call and perceive as existence. Too much Being is not what we are, too much Nothingness likewise ceases to be what makes us what we are, and yet, for a brief moment that for us corresponds to the totality of experience, balanced on the subtle path, we are.

2.1. The subject in front of nothingness

A further question opens here, a way of being aware of *nothingness*, it is the gap between self and that which is external to the subject, be it the world on one hand or the idea on the other. The *ego* feels an unbridgeable distance between itself and the other than itself, and this infinite denominator of the ego-world relation produces as a result an abyss into which the consciousness often fears to lose itself. It is particularly difficult to fully understand how a subjectivity can come to be identified with an idea, because the thinker is necessarily "that individual", or else the condition of possibility that there is the subject and thus the thinker himself is in danger of disappearing; the ego, mind or subject tries to follow this path in an attempt to escape from Nothingness, to try to find a solution to the abyss that existence reveals to him.

Schopenhauer is probably also influenced on this issue by oriental suggestions, which may escape our sensitivity. Even a fine hermeneut like Maurizio Malaguti declared with candour its difficulty in fully understanding this possibility of the subject identifying with the idea. Nevertheless, the tension that this thought provokes is powerful and sustains thought. Reflecting on Nothingness and the infinite is a search around the principle as that from which subject and object (interiority and exteriority) originate, and it is that *in which* we are thinking. Is the principle in its first object articulation determined as idea (from Plato to Schopenhauer), which are modes of the principle where subject and object are the same? Or does the principle give rise to thinking singularities that are ontologically "mindful" of a more intense and original singularity? The answer is not given to us, nor perhaps will it ever be in the definitive form we seek, but the *idea* is once again the strategy we seek in order to escape Nothingness, to avoid falling prey to it, to prevent the "nihilification" of nothingness from nullifying the being of the beingness'.

In Greek thought, the focus goes to the *nous*: it is the mind that is such because it is mindful of and assimilated to the idea, it recognises it in the things of experience.

Plato describes the *One who* is not the most abstract, but the supreme entity and it is no wonder, given the inadequacy of the mind, that it is experienced by us as the void. The very thing from which Nothingness arises to the human mind as antithesis appears to us as that which is closest to it. But at the same time, the same search for meaning leaves the way open for Schopenhauer, who draws opposing images from it; yet its obscurity does not contradict the singularity thematised by Plato, and indeed supports its understanding, summarises it and even completes it, leaving each person free in the highest sense to direct his or her own mind to the search. "To think" is to think Being as principle and condition of possibility, and this is the role of philosophy. Not to give answers and least of all to defend them, but to ask questions and support the exploration of what cannot be conquered but only recalled upon returning to oneself.

3. The Nothing, between history and speculation

As is often the case, questions open up paths far more intricate than those imagined at the moment they are posed. And this is probably how, at least in the West, reflection on the theme of Nothingness began. In the Greek world, politically among the freest of antiquity from the perspectives and answers of religious inspiration, when philosophy, and thus reason, emancipated itself from the mythical dimension and the imaginative customs of oral tradition, the question about the origin of man and the world uncovered a Pandora's box.

"Where do we come from?" is the formula with which this eternal question has been handed down to us by tradition and to which the most varied answers have been given. However, they can be summarised in two large families: either we come from somewhere, or we come from nowhere. Of the options, it goes without saying that the first is the one on which our tradition has focused with the greatest interest, and myths abound, first, and then theories, which seek to identify a principle and justify the generating mechanism of the reality we know. The second option, on the other hand, appears almost self-contradictory.

and before that a mockery. How is it possible that the whole, reality, originates from Nothing? We shall see in a later section what difficulties, primarily logical, need to be borne in mind when attempting to answer this order of questions; for now, however, let us continue by sticking to the central theme of this chapter: Nothingness, and the role it has played in the tradition to which we are heirs.

While answering the question of origin in an absolute sense is the challenge we are talking about, and perhaps cannot by nature have an answer that satisfies us without requiring a certain amount of reliance on the part of the recipient, it is instead possible and far easier to answer the question of the origin of defined entities.

Where does "nothing", as a term, come from? In English it is *nothing*, literally not-thing, no thing, in French *rien*, from *rem*, accusative of the Latin *res*, the reference to "thing" is explicit, going all the way back to Latin *nihil, ni-hilo* i.e. not even a hair, not even the slightest thing. The Greek, *ouden* is a more intricate philological challenge, and in the reading of Rocci – a historical dictionary compiled by the Jesuit Greek scholar of the same name over 25 years of work – it appears as the progenitor of all previous translations, combining the prefix *ou-*, which indicates negation, with *-den*, which in Rocci is associated with the Latin *quidem*. However, this is not the only way of understanding the term, which up to this point is much more akin to the word "nothing" than to the word "nothing". It is the Greek that offers us an opportunity to shed light on the matter, since the pre-fixed *ou-* in addition to negation is also used with an adversative function, as a "but". And so to the more pragmatic version of this term, and of which we are heirs without any doubt, transposing Nothingness into our Western culture as the negation of being, as if it were an absence or an erasure, the early Greek may have attempted to point to a more complex perspective – and perhaps for this reason less successful in language, which benefits from simplicity and clarity by being intended for use – in which Nothingness is posited as an alternative, not so much to being as to Being. Such a perspective is first considered by Parmenides; as Prof. Cantalupi will also recall in the following pages. The philosopher from Elea can be considered the author who sets the question by making Nothingness a ni-entity, and attributing to the latter a character intrinsically contradictory, since as the negation of something that is, i.e. as "not" with respect to an entity, non-being presupposes Being in its conception in an insuperable way. The very expression "not being" is derived from, and needs reference to, Being in order for it to be expressed and understood. In this way, Nothingness, reduced to "ni-ente", loses its autonomy, and no longer refers to a particular concept or thought, but ends up representing a simple negation.

The question is not one that meekly submits to the cathegories of the human intellect. The incommensurability of Being and

Nothingness eludes our experience, and logic, although it may be a valuable compass, often does not involve the totality of our mind, leaving us as in possession of well-made but unsatisfactory answers like a cold meal after a day in the rain.

Before delving, then, into this rigorous but sometimes slightly alien dimension, let us again ask the Greek world to give body and warmth to our quest. Not exactly a metaphor for nothingness (*ouden*), nobody (*oudeys*) is a related term both in construction and in the intention of its reference, aimed at designating a void.

The first and most universal of the authentically mythical tales is the Odyssey, and it is from this place of the soul that we want to take our cue,

in particular from the story of the Cyclops Polyphemus. Most of us are at least vaguely familiar with the vicissitude of Odysseus, who falls prisoner to the giant and presents himself to him as Nobody. No translation can however render the author's mastery, for the hero's original name, Odysseus, is almost homophonic with *oudeys*. Thus the famous "my name is Nobody", reveals a depth even greater than that which already connotes its author in the imagination of us all. He can only declare himself to be Nobody, and thus embody a contradiction, by exploiting a blurring between language and concept. He does not lie Odysseus, not really at least. Nor does he tell the truth. He stands poised between Being and non-being, in that middle way denounced by Parmenides, and of which, however, not enough fragments have reached us to be able to interpret it fully. Thus Nothingness resembles non-being, nothingness, and certainly in everyday life the two expressions are comfortably superimposable.

But that an expedient works; that a simplification satisfies, does not mean that it achieves the objective, especially if this is on a theoretical level, and cannot therefore be exhausted in a practical dimension. Induction, i.e. the generalisation of particular and determinate experiences by repeated abstractions, cannot reach the absolute. Perhaps the absolute simply cannot be reached, but to test this option, we refer, as mentioned, to a later reflection.

We are determined by the structure and functioning dynamics of our cognitive apparatus, which cannot conceive absence except as negation. Maurizio Malaguti used to exemplify the issue by pointing to the chair and saying that there is no flower vase above it; and immediately preceding to ask if we had not imagined it at the very moment when its non-existence was declared. A similar reference has become known for a popular work on the subject of communication, in which the reader is urged not to think of a white bear. Inevitably, the image appears in everyone's mind.

What is crucial in order to understand the question of Nothingness and Being is that it is a metaphysical problem, which constitutively goes beyond the threshold of our experience, and which cannot be answered in the form that habit offers us for everyday matters. Attempting to grasp the specificity of the question requires us to abstract ourselves as far as possible from the determinacy and particularity of our perspective, looking at the same time in our deepest intimacy and beyond ourselves. Both Nothingness and Being are not experiences among experiences, but something qualitatively other, on which experiencing, feeling and perhaps even reasoning depend, but which they are not likely to exhaust.

Odysseus can blind Polyphemus because he has only one eye; he would have had no time, resources or manpower to hit two eyes at the same time. However sharp his eye is, it is incapable of perspective, it cannot grasp the blur and is blinded by this inability to grasp the multiplicity and complexity of reality. The meaning of Polyphemus's name is the note that contributes to its eternal value: Polyphemus is literally "he who speaks much". As in all times happens to those who see the world in one light, doubt and reflection do not punctuate the words, which flow incessantly but inconsistently. To the point that the Cyclops, wounded and injured, shouting his wrath and crying out for help to his brothers, is not even able to communicate with them; he cannot stop his stream of consciousness to clarify it to others, because he is blind to what is not unambiguous and cannot get out of himself. Thus the hero of multifaceted ingenuity can regain his freedom and return to being lost in the world.

3.1 Nothingness under the lens of logic

The first author to question the subject of Being is Democritus of Abdera, Leucippus pupil. A decade older than Socrates, he is canonically counted among pre-Socratics for his "physical" research for the foundation of reality. His invention of the atom as nature's prime element should not be confused by its assonance with the language of modern science. In fact, Democritus' atoms are indivisible elements (the very term a-tome literally means indivisible), which have all the characteristics of the Parmenidian Being, with the exception of uniqueness. Atoms are, in a sense, Being declined in a plural way by the philosopher of Abdera. However, this difference is sufficient to unhinge the very structure of classical thought. On one hand, it will produce that current of thought that will slowly make its way under the name of mechanism, marking the fate of the modern era and representing the foundation on which scientific thought will take root. On the other hand, the democritean system, for its own coherence, requires the invention – at least in the Western context – of the vacuum. Specifically, its inventor derived the concept from the need for a space in which atoms could move, and which was therefore empty.

The opposition for Democritus is thus between fullness and emptiness, and not necessarily between being and nothingness. However, the concept of emptiness, made absolute in the objections of the Eleatic school on the subject – which, simplifying the matter to a great extent, we can consider to be at the origin of rationalism – takes on that metaphysical hue that brings it closer to nothingness. Nothingness and emptiness have since then been intertwined in the history of thought, generating many significant plane confusions that are still difficult to untangle today. This original idea is completely foreign to the culture and modes of thought that had hitherto characterised the Mediterranean basin, but from that moment on, it will never again leave the West, and no philosophy or religion will be able to escape confrontation on this issue. What is striking, and what could still be an important warning today, is that the victory of the Parmenidean thesis is a capital question for our science, which has "delayed" almost two thousand years to evolve, not being able to formulate the notion of "0" in mathematics and not having been able to embrace that of "vacuum" in physics. This is a metaphysical question, which probably arose from the linguistic complication of clearly designating what was meant by the term "negative".

This nothingness, which perhaps stems – if you will pardon the pun – from "nothingness", from the negation of something, and which, as we said at the beginning of this chapter, has thus also been received by modern languages, designates, or at least attempts to designate, a completely different order of reality. On the one hand, nothingness, and thus the concept that we originally attempted to designate with emptiness, is an absence, a space of possibility that is open and not yet determined; its counterpart is what is obtained by removing the negation [ni – entity], and thus entity: any element of reality that is definable and circumscribable at least on the communicative level. This metaphysically minimal position, which suspends judgement and does not delve into the question of the nature of entities and the void but limits itself to naming them effectively on the level of linguistic pragmatics, stops at the possibility of generalising nothingness and the thing. Taking this step, one instead enters the realm of Nothingness and Being. The greatest confusion and difficulty in understanding arises when it is not clear on which of the two planes one is moving, and with what purpose one is constructing arguments. This is probably what happened between Democritus and Parmenides. The latter, applying the logical principle of non-contradiction – whereby one cannot deny and affirm at the same time the same predicate of the same subject – he thought he had demonstrated the contradictory nature of nothingness as an entity. Instead, he had come to anticipate Kant's thesis of the inconsistency of being as predicate. Even Kant, however, failed to grasp that this thesis of his is incomplete, and that his demonstration is "limited" to predicates of the first order. That is, that it cannot be said or denied that a thing is in absolute, but only that it is or is not in a certain way. This further passage was to be the source of numerous paradoxes, some of which have not yet been unambiguously resolved and are the basis of philosophical (and even scientific) positions that continue to be opposed to this day.

At this point in our discussion, it may perhaps help to anticipate a philosophical position known under the terms of *conceptual natural realism,* and which compared to the traditional classifications that take part in the dispute on universals lies somewhere in the ranks of the moderate realists. Today, many sciences from more than one point of view have revealed how an objective view of reality is perhaps elusive for our knowledge, and that the best way to get something to help one orientate oneself in the world can be achieved by stating the presuppositions (or prejudices) from which each perspective approaches the world and to what ends it is orientated. However, in order to understand Being and its complementary, Nothingness, which we have outlined as two thoughts that have emerged from attempts to generalise to the absolute of emptiness and being, in a new and paradoxically more authentic perspective, the West had to wait until Plotinus. It's the Neo-Platonic author who, in his *Enneads,* reasons about Nothingness and defines an ontological statute, so to speak out of the apparent paradox. Both Being and Nothingness are approached by the Neo-Platonic author by following the path of abstraction. At first glance, the directions appear opposite, but both the One – as the foundation and metaphysical principle of reality is called in his thought – and Nothingness coincide and constitute the essence of things. These "two species of nothingness", one above, the other below, limit every possible entity. In particular, the Being-One is defined as a nothingness with respect to the entity, due to the fact that it can be indicated only in the negative.

It will be following this argumentative style, combined with the intention to applicating reflection on an originally logical theme such as the one concerned to questions about the nature of the Christian God, which will usher in the season of negative theology (elsewhere in these pages we will refer to the apophatic way of searching for Being to qualify this mode of theoretical research). It is the reflection on God – assumed as the *alter ego* of Being in a cultural landscape with Christian traction – which takes as its pre-assumption its radical otherness with respect to anything determined.

Thus God, totally other than entities, transcending any possible experience, is conceived as nothing compared to the phenomena with which we can relate. The logic behind this interpretation presupposes a negative definition of being: everything is a negation of everything else; the so-called *principium individuationis* (which came back into vogue in the 19th century with Schopenhauer) is what makes each being what it is *and nothing else*, thus defining the individual as a limit to any form of otherness. A = A, i.e. A = *not not* A. God-Being is totally other than any entity, hence the radical negation of any particular individuality, it is not a defined "that", but "that in which" the definition can take place, and which transcends them all; in this sense it is the negation of entity. God-Being is thus absolute affirmation because it is negation of the negation that is being, and the double negation is logically equivalent to an affirmation.

In the 15th century, this path would come together in Nicola Cusano's *De non aliud,* in which God is referred to as the *non-other*, that which unites everything and is thus distinguished from everything else.

The metaphysical question of Being is by this time a theological fact, and the implications of this, which without controversy we could consider as a drift from the originality of the question, will need a few centuries to return to a more directly logical relevance.

It was Kurt Gödel who, in the first half of the 20th century, put the matter in order by showing how Being and consequently Nothingness, as absolutes, are intrinsically inaccessible, or with what limits one can attempt to refer to them.

The story of this character would deserve a biographical treatment in its own right, for the genius that made him live a life on the threshold of a world of *unsustainable lightness*. Gödel was a mathematician, and in particular a logician, among the greatest in history along with Frege and Aristotle. In the pages that follow, and it is important to point this out at the outset, we will deal with logic and physics, but without any claim to completeness and softening the formalisms for the purposes of a discussion that is intended to be theoretically oriented. We would therefore like to avoid the objection of paralogism to the argumentative structure by anticipating for the more specialised reader that the following structure is intended to serve as a metaphor, useful for the expression and dissemination of a question – that of Being and Nothingness – which by its very nature has eluded any attempt of definition for 25 centuries.

What is important to the writer is to offer an insight into the problem, its context in the current landscape, and to compare it with what our scientific paradigm currently allows us to say about the nature of which we are a part. So let us begin this stage of our journey by reminding ourselves that science teaches us that no certainty can claim absoluteness, and that definitions are convictions, and that they age the better they are prepared to embrace their own becoming.

The young Kurt took his first steps by tackling the most ambitious challenge in modern mathematics: Hilbert's programme. This initiative, undertaken by the great Austrian mathematician and handed over to the scientific community almost as a spiritual mission, was aimed at proving that mathematics, starting from a finite number of premises (axioms), was contradictory. Gödel, who was only 24 years old, presented a thesis, which was later proven by many, and which we know today as the Incompleteness Theorem. In a nutshell, it states that any axiomatic system, such as mathematics, for example, cannot prove both its consistency and coherence using only its constituent elements. A step back to recall the meaning in context of the key terms. The word *axiom* refers to any formalisation of the properties that constitute the definition of the object of the theory. We can think of them as the minimum and essential structure for a given object to exist. We add that in more practical fields, such as the physical sciences, a role of axiom can also be taken to mean those propositions to which a self-evident truth is attributed (a term we will not dwell on, as this would open another chapter at least as thorny as the present one), something that is indispensable in order to orient oneself in the phenomena one is trying to move between, and which one aspires to relate to one another. *Coherence* is then the way to qualify a formal logical theory within which it is not possible to prove a contradiction; that is, a theory is coherent if, starting from its axioms – to paraphrase another immense logician like Wittgenstein – if, playing by the rules of the theory, it is not possible to prove both a statement and its contrary. Finally, a theory is said to be *complete* if its axioms, i.e., its foundations, are sufficient so that one can distinguish between a true and a false statement within the theory itself. Gödel showed that within any axiomatic system powerful enough to construct the integers cannot be both consistent and complete.

The implications were disruptive, but the author himself did not judge it to be a tombstone laid on mathematical realism,

Gödel himself, on the other hand, interpreted his argument as a confirmation of the Platonic approach (progenitor of all realism), and as a clarification of the fact that truth, as an objective, cannot be subsumed in any context, not even within the confines of demonstration.

Returning to the specificity of the Incompleteness Theorem, it decrees in practice that what is possible to say within a coherent axiomatized system is more than what is possible to prove within the same boundaries. The consequence is that if a certain system is coherent, it is impossible to prove its non-contradiction with the resources offered by the system itself.

Let us now turn to the point that touches us most closely: what does the "interior" of a theory or axiomatic system represent for us? This question can perhaps be clarified if we take a step from mathematics to physics, bearing in mind that the former is the language of the latter. Physics i.e. constructs its descriptions of the world and the relations between the phenomena it examines by using mathematical relations between variables to which it corresponds physical quantities, i.e. defined and measurable properties in the world around us. Applying then what Gödel demonstrated – and knowing that what applies to formal logic cannot simply be transferred to the special physics ontology – the natural sciences cannot in the first instance demonstrate the origin of nature itself. In this sense, the Big Bang would be a description of a natural phenomenon, but it does not answer the question of the origin of the Universe; it answers the question of *how it* came into being, but not the question of where it came from. For this purpose, other orders of thought arise, attempting to account for those statements that are indemonstrable within the constraints of a deductive theory, but which can – perhaps – be resolved by using a metalanguage. That is, a further level of discourse, a system of rules that contains and defines what in the previous context remained undecidable. A criticism of this strategy is that it does not solve problems, but merely shifts them a little further, inventing solutions as the rules of the new reference system within which the previous one is included are posed and being able in principle to proceed in this way ad infinitum. This, however, is not the intent with which metaphysics, for example, was born, that reflection that in the intention of the first thinkers attempted to answer those questions that the order of phenomena would have in their eyes – and still in ours – left without a sure answer. Rather, metaphysics aims to define those undecidabilities that the analysis of reality reveals, and on these areas, it tries to exercise that spark of a very special nature that is the human mind in all its manifestations, from art to science, but knowing that the resulting discourse cannot be a description of an object, but an announcement, a narration. Being and Nothingness, later also God, appear as those parts of nature whose need can be felt, whose concept can be attempted, yet precluded from experience.

Infinity, which more than Being is the object of mathematical analysis, more than an object in itself shows itself as the behaviour of certain objects within a theory. To a practical end – assuming one can refer to such an object with this term – designates an incomparably large quantity compared to others being considered; it can also sometimes be used to refer to the whole, but this second option, frequent in everyday language, is not adopted in the measured lexicon of logic and philosophy. Nothingness, on the other hand, has its own more specific identity in this sense, and something similar, which is the void, enters the analyses of science. We have already alluded to this term and its appearance on the scene of physics, and it is now time to go into more detail, to understand its importance but also its – literally essential – difference from Nothingness.

3.2 Almost friends: Nothingness and emptiness

Nothingness, we have attempted to express it, is a term used to denote an idea belonging to the domain of metaphysics; it indicates the absolute opposite of Being. It is not an entity, nor is it an attribute that some object of nature might manifest in our eyes. The story of emptiness, however, is a different one, starting with the classical reflections of Leucippus and Democritus – to which we have already declared our deep intellectual debt – and meandering like a karst river up to modernity, when the scientific method gradually removed the ideological encrustations and made it possible to look at nature without the lens of dogmatism. It is then that emptiness returns to the scene, becoming a major player in many fields.

It is in Newton's *Principia* (1687) that the void re-emerges, after centuries under trace, understood by the English physicist as the container of material objects; in clear coherence with the democritean vision that made it the space of atom's action. It was later to be Kant, who played a major role in this, who interpreted Newtonian space in the *Critique of Pure Reason* as an a *priori intuition*, which, together with time, constitutes the load-bearing architecture of human knowledge. It is still an absolute that is hypostatised, and precisely by the author who contributed to recognising this same type of work as unproductive. It does, however, have the merit of re-establishing that partnership between philosophy and physics that had been at the origin of classical thought but had faded until modernity.

It would take little more than a century for this view too to be further overtaken. First in a non-substantial way, with Einstein's Narrow Relativity, which thinks of a four-dimensional continuum, still empty and static like Newtonian but no longer Euclidean. Whereas in General Relativity, matter takes on the appearance of a discontinuity in the gravitational field.

Emptiness (space) is recognised as the tendency to dilute material structures that lack sufficient mass to change their geometry. Just as it is said of love, which either grows or fades but cannot remain motionless, with this metaphor material entities (but perhaps not only) cannot remain eternally stable in space-time, but either become centres of gravity, mathematically attractors, for material dynamics external to them but which, due to the effect they exert on the form of space, will tend to converge there; or they disintegrate and their components lose that cohesion that made them that defined entity.

It is worth pausing for a moment before going any further and clarifying what is meant by the term *"entity"*. A term that is by no means taken for granted, but perhaps a little less elusive than Being and Nothingness, at least for the fact that *entities* are all the objects we habitually frequent, that we ourselves are and from which it is constitutively impossible to escape: except (perhaps) by ceasing to exist.

We then propose to think of the entity as *the metastable node of a relationship between the individual and the environment*. That is, as a portion of reality, a confluence of matter and information, which manifests characteristics of cohesion and coherence with respect to a given reference system. A piece of nature that acts as if its parts were a unit in a given context. The term "metastability" is intended to indicate the temporariness, and more generally the relativity-contextuality of this characteristic. Entities, as we know, are not eternal. Entity is thus a mode of being in relation to. This view is borrowed from *children's physics*, as quantum physics was called in its gestation years, since it was invented by four very young scientists whose ideas have turned the world upside down in ways that have not yet fully unfolded. The void within this new paradigm is maximised by the possibility of virtual particle pairs, floating fields, manifesting in it. The result is that every entity is conceivable as an actualisation (act in the words of Aristotle) of an infinite possibility, the constraint of which is described by the *uncertainty principle* formulated by Heisenberg. It is thus admissible that matter (or energy, which is the same thing as Einstein demonstrated in his most famous equation) manifests itself from the vacuum (which is not nothingness) without violating the principle of conservation of energy – which in short states that nothing is created, nothing is destroyed – since this principle states that the product of the variation in energy and the measurable persistence of an event is greater than Planck's constant (canonically, h).

In practice, any amount of energy can be used "in debt" to the universe, provided that the product between this and the "restitution" time is less than h. Entities are then conceivable as those places and at the same time moments – an equivalence once again established by Einstein – in which the vacuum instantiates itself in one of the possible ways. This is also referred to as *quantum field theory* (QFT).

The stability of entities, which we have proposed to define by the term *metastability*, is a direct and insuperable consequence of the principle of uncertainty. The fact of arising from the void and being destined to return to it is likewise a consequence of this principle, and a thought that was formulated by Anaximander at the origin of classicism. The beginnings of science and the most avant-garde modernity draw closer and perhaps even recognise each other.

4. From where?

If then we are given neither knowledge of Nothingness nor of Being, just as we know about the things that surround us and that we ourselves are, simply because they are not things. In fact, naming them is not enough for them to become substances, and there were more confusions and vagueness along this path than the certainties that it was possible to acquire. Physics, both ancient and even more modern, does not satisfy us in explaining nature and its mechanisms; they do not clarify it for us.

The quantum vacuum and field theories impose on our mind concepts that escape the representative capacity, confuse more than they orient, and offer the possibility of numerous and picturesque misunderstandings. Thus, the most disparate thoughts legitimize themselves, seasoning with the term "quantum" any mental chimera that claims to mimic metaphysics. There is a single common trait: the desire to satisfy a spiritual longing to go up the river of existence to its source. Another leap back to the dawn of civilizations reveals a common element in the conceptions of the world between cultures that have never faced each other. Ancient Babylon and the Egypt of the pharaohs, the Maya and the people who arose around the Yellow River, having only the sky to look at in common, all developed a cyclical vision of time and nature, combined with a good dose of fatalism and an immanentist conception of divinity.

Without going into an archaeological expedition on the term *psike*, with which the Hebrew word *nefesh* is translated, and from whose semantic confluence the modern conception of the soul was born with all that comes from it, but assuming that the transcendence of the principle is an intuition-invention of Plato, we must keep in mind that the thousands of years that separate the Greek philosopher and the consolidation of his thought in the West are preceded by millennia of immanentism or pantheism. That principle (or Creator) which Christianity leads even the most convinced atheists out of habit to think of as separate and independent from its creation, was – and in many Eastern cultures especially still is – coincident with nature itself, uncreated and eternal in become. A cycle of beings (entities) from which one cannot escape, not even to peek just beyond the center of this vortex. Simply because the center or origin is unknown.

The theme of the origin of the cosmos that we have outlined in the previous pages, both on the physical level, which, as a question profoundly linked to the logical dimension and the limits that reason must recognize with respect to its own products, also arises in the form of eternal problem on the nature of time. Thus, the concept of eternity, liminal with respect to the question, offers two distinct interpretations: one, more immediate in appearance, in which the eternal is conceived as something that extends infinitely into the past and the future. In this way, however, the problem shifts; in fact, what is infinity? What does it represent with respect to the insurmountable finitude which is the very figure of the entity as such?

Useful in mathematical analysis, infinity is however a property that existence does not grasp, and which, in defiance of the habit of using it as a hyperbole or as a metaphor, does not designate knowable properties of the entities that are offered to us on the level of existence.

Another way of relating to the question is that taken by Augustine of Hippo, who thinks of the eternal as "other than the passing of time"; in this way Creation would not happen in time, but with time as he himself writes.

So, is seeking the origin of the universe more of a mind's need that contemplates nature while sensing its own existence, than a logical necessity of the cosmos as such? The writer thinks, indeed he is certain, that seeking an answer to the origin of the cosmos is an "objective necessity", and a logical necessity that the inhabitants of the universe cannot fail to consider. And that in the final analysis "something" could have appeared from the "void", as a real entity or as the memory of a becoming and transitory self, but capable of thinking its own beginning, and in this of pushing thought to remember the beginning of all the "plots", looking for a first node capable of holding each plot together, represents a possibility that is anything but remote.

CHAPTER II

THE UNIVERSE

Universe, Cosmos, perhaps even Nature, are among the names most often used to try to encapsulate the immensity of all that surrounds us. Before we embark together on this second section of our journey through Nothingness, the Universe and the Infinite, let us pause for just a moment to take in the panorama into which we are venturing. Before us stands – literally – everything. Every atom and every galaxy. Every wave, ray of light or sound, every combination of matter, every thought and every dream. All of this is the Universe. And how is it possible, or what sense can it make, to ask questions about something so immense and heterogeneous? This is perhaps the answer for another book that would dare to venture into the folds of that particular part of nature that is the human soul. As far as we are concerned, here we content ourselves with picking up a baton, acknowledging and accepting questions about the world we are part of as a given, almost as if it were an instinct from which we do not want to (or cannot) escape. Certainly, pointing our gaze into the borderlands arouses some shivers, a sense of both awe and wonder, of respect and vertigo. And we want to welcome it, not with the audacity to win a challenge, but driven by the desire to feel part of this enchantment; we take courage, then, with the words of an unparalleled icon of human adventure: Marie Sklodowska Curie, who reminds us that in this life nothing is to be feared but only to be understood, and it is precisely because of less fear that we continue on our path of research.

1. What is the Universe?

It seems easy to answer: all that there is. But what does it mean "to be there"? Venturing into the heart of our journey, impossible questions arise again, but let us not be led astray. What we intend to examine here as the object of our research is that specific mode of being that characterises things we call concrete; those we instinctively call "things", to be clear.

Alongside this Universe of objects of all shapes and sizes, there are at least two other orders: one is our thoughts, the objects of all those insuperably individual experiences that occur within our consciousness; this is our personal Universe, in which dreams and feelings, emotions and ideas find a home, which we can express in the world as if by creating external duplicates of it, but which never fully reflect the immediacy of our inner feeling. Among these universes of consciousness, there is one particular type that is an exception in this latter respect: mathematical-logical objects. These, in fact, show themselves in the world of things as endowed with an objectivity and univocity that makes them supremely effective in communicating the characteristics of the material nature in which we are immersed down to the most minute details. By contrast, they do not apply to any extent to the other objects of consciousness, which for their part radically elude any attempt at quantification.

What we call the Universe seems to be distinguishable into at least into three other ways, which comprises what is the material nature for us, what belongs to consciousness and what seems to come from that further realm that is the world of mathematical ideas (and which sounds so much like Platonism). In the pages that follow, we will focus especially on the first aspect, and references to the other two will be stated and circumscribed, in an attempt to keep the path orderly and to explicate the off-track shortcuts that we are sometimes obliged to take in order to overcome the most massive obstacles.

If we refer to the Universe with the first meaning, we place ourselves in the groove of Western tradition. The idea of conceiving Nature as that immutable background that no god and no man made, is characteristic of Ancient Greece. Just as Greek is the term cosmos, which designates a unitary and harmonious whole. For the man of the ancient world, the nature in which we live is a given, something rhythmic and eternal in its foundation. Human possibilities in relating to this dimension are primarily contemplative, in the sense that nature can be known but not possessed; it is conceived as something in which man is constituted and not as a gift, a bequest. The perspective changes with the advent and spread of Judeo-Christian thought that transforms the world into "Creation" and man its lord.

This cultural approach was to hold sway for many centuries, and even today its trace in the way we relate to nature has powerful and extremely relevant implications; but this is not the topic we are interested in, we will simply consider that this perspective in no way interests us at this early stage. Instead, our aim will be to come as close as possible to something like an objective description (and we will clarify whether and how this can be done with regard to this "object") of the Universe as cosmos or nature, attempting to recover, in the light of the possibilities and knowledge that have matured over the centuries, the same cognitive intention as the ancients and to confront nature with a rational gaze. In this sense, words come to our aid, and the object of our research is perhaps closest to the Universe understood in the manner of astronomy rather than words like "cosmos" or "nature".

At this point, a number of questions come into play, some very serious, others that seem like word games, and still others that arise from the fact that the words we use in formulating them are the same as those we use on a daily basis, but which in scientific language denote elements that differ from those of common experience. For example, an intuitive definition of the Universe might resemble: all that there is. As we have already said, "being" is a complicated concept, but here let us limit ourselves to using this word in a non-philosophical way as a synonym for existing, and let's focus instead on the term "everything". What does everything mean? We can think of the whole as the content of the whole that contains every object (every entity) that currently exists. We have then introduced a further complexification: the time scale. And this legitimately opens up questions about the location of the past. In two steps, attempting to look with just a little more precision at our object of investigation, we fell into set theory (already a cardinal element of this book) and the space-time relation. Because instinctively for us, if we don't think about it too much and answer straight away, the Universe is a place. The place where all the things we can see with our means of investigation are. Here again an ensemble instinct peeps in, immediately followed by the question: but where is this place (this ensemble)? The astrophysical answer is that the Universe is not somewhere, is not itself contained in some dimension, and the notions of space and even time only make sense within the Universe and not for the Universe.

1.2. Towards infinity, but beyond?

Thus, another instinctive question arises: but if the Universe was born with the Big Bang, was there no before? And once again, the official answer comes back as ungratifying: there is no before, because time (as well as space, since from Einstein onwards they seem to be sides of the same coin) comes into being with the Big Bang, while the Big Bang does – not take place in time. In this sense, the Big Bang is the furthest place/moment from us of which we can experience. Beyond that, the notions of time and space, and consequently those of cause and effect, cease to have meaning.

We can try to understand a little more clearly what this means by striving not to think – contrary to our habit – of the Universe as a thing among things, but as a property of all things. A way of being rather than a container. And, as the name itself suggests, it is the common direction, the one direction in fact, in which all things we know are oriented. The fact that they are in relation to each other, that they respond to the same laws (or so it seems from our vantage point. To use a somewhat more obscure expression, the Universe is the possible causal reciprocity of a collection of entities, and that is why we, who are part of this collection, conceive of them as existing. In this way, we have constructed a kind of snapshot of all that there is. Let us keep the reference to everything on the bench for a moment longer, with the promise that shortly we will – hopefully – illuminate the matter just a little more.

Some entities, most of them actually, cross time; that is, they have what we call a duration. It means that when we look at a certain snapshot, that is from a given moment, we expect them to be in the next snapshot, or to be in the one before that. A human being, for example, is in the picture of the Unverse for about eighty years; a star for billions of years. The idea is quite simple if we forgive ourselves a little approximation, as for example if we imagine that we take a photograph in a certain geographical co-ordinate; where we see Mont Blanc today and can expect to find it in the same position relative to the planet earth in the same shot 10,000 years ago, if we were to take images from 10 million years back in time from our album, we would probably see a completely different landscape.

By tying the coordinates to our planet, we have relativised the position, "forgetting" time in a certain sense. It is easy to understand this if we instead think of keeping our position fixed not with respect to the earth's latitude and longitude but with respect to our sun; in that case, after a few minutes we would find ourselves taking a picture while floating in space and after six months the earth would appear as a blue dot at the antipodes of orbit.

Hopefully, we have thus rendered the idea of how what we consider a "moment" and what we conceive of as a "place" are actually two inseparable aspects of an event, a point defined together by three spatial and temporal dimensions. There is no *here* without a *now*, nor a *now* without a *here*.

Here we can now return to the subject of "everything" and attempt to clarify in a more precise and measurable way what falls under this label. Let us take as an example a certain point in space-time, which is fine for the dimensions to which we will refer to your *now*. From there it can only have an effect on you, and thus can only exist for you, whatever reaches you.

In principle, any event in the past can, with enough time, reach any point in the future. Reach in the sense that a photon (the component of sunlight rays) can start from the past event and survive until it reaches a certain point in the future. And so energy and information are transmitted in a causal chain on which the current structure of that portion of the Universe depends. The way in which this form of the space-time is mathematically represented is called the light cone, imagining an hourglass form in which the present is the centre in which the two cones join, where the "lower" cone represents the set of past events causally related to the present, and the "upper" cone represents the set of possible (accessible) events from the here-and-now.

Light and its speed are the limit of what can be "present"; for example, the rising or setting of the sun reaches us about 8 minutes later than when it peeps over the horizon line because light takes that long to travel the space between us and our star. And so, if the sun disappeared in an instant as by magic, we would notice it 8 minutes later. The question of what the whole is and consequently what the Universe is depends on how light works.

Thus, only one and final piece is missing to complete our puzzle and precisely circumscribe our Universe; truly *ours*. In fact, if things were only in this way, without further elements which would complicate the question, the present would be definable as a set of causes and information that could reach a certain point (a certain observer) at a given instant, and it would be relatively simple to calculate the extension of space-time from which these came from. Theoretically in this case the Universe reachable to us, with sufficient time and patience available, would be unlimited. As you can imagine, however, nature is much more fascinating and mysterious than we would like.

The Universe in fact, far from being the unchanging background that the ancient Greeks thought, is expanding. The story of how we came to realise this is an astonishing page of great 20th century science that unhinged another of the cornerstones of our worldview, showing once again how reality is often able to surpass imagination. Indeed, as soon as we are confronted with this argument – supported by the empirical evidence that all galaxies are moving away from us, and moreover at an increasing speed directly proportional to their distance – questions arise such as: where is the centre of this expansion? Where is the Universe expanding? What is outside? Let us see, then, what is meant by the fact that space expands, and having clarified this, we trust that the relationship with what was said earlier about the role of light will give us a clearer and more objective view of what we mean by the Universe.

Since the Universe is not one thing among others, as we have repeatedly attempted to show, its expansion is not to be understood as imagining an oil slick gaining, instant by instant, ground on a given surface. If only because there is nothing else beyond the Universe and, recalling another theme that is at the heart of the reflections in these pages, "nothing" in the absolute sense can perhaps only properly refer to that which exceeds the Universe. The Universe expands not in something but in itself, in the sense that the amount of space increases with the passage of time; as if on a sheet of squared paper, as a certain amount of time passes, a new square appears among those that were already on the sheet. So, if we colour two adjoining squares, after a certain amount of time, they will be separated by a new white square, and so on. If instead of the coloured squares we think of galaxies in our Universe, that's how we would justify the increase of expansion and speed. In fact, if a third one appears between each pair of squares after a given time interval, and the two previously contiguous coloured squares are now separated, after a new time interval, the same will happen between each of the two-coloured squares and the new one that has come between them, thus being three squares apart. After a further interval of time between each pair of squares, a further one will appear, and our initially coloured squares will be seven squares apart. This also explains why the expansion of the Universe is so much greater the further apart the ends of this expansion are and why it appears as accelerated.

We finally have all the elements to accurately describe what the Universe is, at least as understood by astronomers. In fact, comparing the speed of light (about 300,000 km/s), which corresponds, as we have already seen, to the limit speed at which two events can form a relationship with each other, and the speed of expansion of the Universe (74.03 km/s per Megaparsec. 1 Megaparsec is equivalent to 3.3 million light years) the difference between the two allows us to calculate the limit of the Universe observable in principle and corresponds to 13.7 billion light years. This is because two objects at a greater distance move away at a greater speed than the speed at which light travels from one to the other, and therefore, mutually, they do not exist. A funny consequence of this is that each of us lives in his own Universe, in the sense that each of us is the centre of a Universe with limited dimensions and therefore also with horizons slightly different.

1.3. The Universe born from nothing and God

The Universe described with the criteria of science is thus surprisingly similar to the Universe thought by "philosophers", in other words what we designate in everyday language when we use the word *reality*. On the meaning of this term, which has been variously addressed in Western culture's history and which significantly in the Middle Ages gave rise to a reflection that has gone down in history under the emblematic name of the "dispute over universals", much has been said and perhaps a millennium and a half of diatribes have at least served to set aside the expectation that a definitive word will be said on the subject.

In order to facilitate an overview of the issue, it is therefore appropriate to briefly outline the main positions. On the one hand, *realism*, a Platonic-inspired conception that lies at the origin of the dualisms of every era; this includes, more or less, all worldviews that distinguish two (or more, as in Plotinus' Neo-Platonism) levels of reality: an original and founding, usually transcendent, separate world, a world beyond this world and on which ours depends, an absolute and necessary dimension that explains the existence of our reality, which is, on the contrary, characterised by transience and finitude. Of this worldview, it is quite intuitive to understand the fortunate alliance with Christianity and the religions of transcendence in general: this world beyond the world satisfies the intuitive and not excessively rational interpretation of notions such as "the kingdom of God" or "the kingdom of heaven" (or what with an almost childlike spontaneity can be called "Paradise"). On the opposite side is *nominalism*, that is, the conception that the rank of existence is a characteristic of those things that "are" in the most immediate and instinctive sense, such as this book or this hand with which you are holding it, while everything else is a mere name; literally air in the teeth, according to the most radical positions such as Roscellino's. To give an example, "white" does not exist in general or in the abstract but is merely the name we give to things that manifest a certain characteristic with respect to our senses. In this reading which collapses on immediacy and immanence there is little place for God or any kind of further reality: ulteriority and transcendence themselves lose their meaning – simplifying a little – due to the fact that they cannot be applied to any entity.

At first sight it might seem that this second option is the closest ancestor to modern science and the current laic conceptions of nature, conceived as a set of interacting forces. But things are not exactly like that; in fact, the so-called laws of nature are not "nameable things", one cannot point them out with the finger whilst naming them, or at least not in a direct manner.

For as an apple falls from a tree, we do not see the law of universal gravitation, but an apple, a tree and the ground on which it falls; what we are referring to when we speak of the laws of nature are not entities or phenomena in themselves, but manifest as (known, regular and measurable) characteristics of entities or phenomena. That is, they are not what Aristotle would have called prime substances, but are more like second substances, that is those that "exist" in entities proper or in the mind of the person who abstracts them from things and attempts a general formulation.

Between the opposite extremes mentioned above, a range of intermediate options has opened up, some close to refined intellectual games and others more fertile. Among the latter is *conceptual natural realism* (we have already encountered it). An intricate name for a theory on the contrary quite simple and close to the intentions of Aristotle, both the father of logic and perhaps the first scholar to whom the title of *scientist* can be attributed with a modern meaning. In this conception, universals, and therefore the Universe itself and Nothingness, extreme concepts such as that of existence or God, just to mention the most pregnant with consequences, are not just names, sounds that come out of our mouths that we use to try to communicate, as the most extreme nominalist positions would have it. Nor, however, do they designate things like the page you are reading. These "entities" exist in you and me as we strive to think of them. In this respect, the law of universal gravitation would be recognisable and recognised as an effective explanation to describe the things around us and to account for their behaviour. In this sense, it would not be a structure or an element of reality, but a – (almost) stable – way in which all things are.

It does not come *before* things, but together with things. It is the way in which things are within a certain horizon of existence, or wanting, of a universe. In the realist-platonic view, we saw this at the beginning of this paragraph, the foundation of the world is constituted by eternal immaterial entities that subsist in a transcendent dimension and the laws of nature thus end up resembling in a certain sense the rules that God would have used to construct the cosmos: a vision that, albeit in a different form, will be taken up again in Part Two of this book and has had some success among mathematicians, who often dwell in this celestial realm with numbers and relations they consider elementary.

The Aristotelian-inspired reading, on the other hand, does not postulate transcendent and ab-solute (that meaning disengaged) entities from a certain type of reality, but takes the world of phenomena as its starting point and abstracts from them those regularities in the mind that we find and which, a posteriori, we call laws.

A question – in the background – prickles us at this point: does a Universe born from Nothing, from that Nothing whose meaning we are striving to discern, given that outlining its boundaries seems a challenge beyond our strength, exclude God? Let me explain: if the hypotheses that will be put forward in Part Two of the book were to identify a "certain" mechanism that would allow us to account for the becoming that has produced what we call the Universe; and if this mechanism were effective even without a premise with those characteristics with which we conceive of God (at least in the manner of monotheistic religions), would this exclude Him as Creator?

This is a question that we are happy to have reached together, but it is not something that can be answered beyond the confines of each of our consciousness. What we hope is that we have just aroused the attention of you who are our companions on this journey. The provocation that Cantalupi advances in the Preface to these pages is a reminder that God, understood in a personal and intentional way, may have *chosen* for the Universe to create itself; He may have chosen for His signature to be indirectly part of His work. Perhaps to let faith be a choice and not a fact.

With these considerations in mind, we retrace our steps for just a moment to take a fresh look at the elusive becoming of our sky.

If a form of life, even one far more skilful and evolved than we could ever be, were to make its appearance in the cosmos in a hundred billion years' time, the Big Bang could not be on its horizon. It would at that time be too far away for any glimpse, for how acute; and who knows what their physics and cosmology could use to fill that gap? For beyond a certain boundary of space-time, things are not part of the same Universe; reciprocally they are not, or if you prefer, they are nothing. The philosophers' Universe, understood as "what is real", can then appear as the set of things to which we are linked by a causal chain. Better: which are linked to each other by a causal chain. So, what is beyond the boundary, behind the event horizon of a black hole or beyond 13.2 billion light years, does not exist. It is not real. It is nothing. This intuition is at the origin of the philosophers' God by Aristotle, who not surprisingly demonstrates its existence as the first link in a causal chain, as the ultimate heaven. The Universe in this sense is what we perceive, not in act, i.e. not that which now, at present, arouses our senses; but in potency, that which by virtue of the very structure of nature is such that it can enter into relation with us, that it is included within the same horizon.

The reason on the basis of physics places any entity beyond this space of possibility as insuperably foreign to us. But that very inaccessible region reveals a gift of metaphysics: if, in that universe beyond our bubble of reality, something like a man looking up at a sky precluded to our eyes ever asks "what am I doing here?", regardless of its form, the laws of its cosmos and the fact that you will never receive the letter of friendship that Carl Sagan sent half a century ago to roam the skies aboard Voyager, that will be the mind of one of our fellow human beings. Physics may tell us that we will never embrace each other, that we will not share the table and other experiences of things that flow, but in seeking the meaning of our existences we will be eternally friends. Not what we can touch, but what we cannot reach and keep trying makes us part of a dimension that cannot exist but is.

The Big Bang is the first act in which the possibility of the relationship is manifested. At first glance, Nothingness might seem to be the impossibility of this actuation; not an object nor a phenomenon, but a name by which to refer to that which cannot enter the realm of entities and phenomena. This extraordinary capacity of ours to catalogue even not accomplishable events by principle under a single name may perhaps have given us the illusion of having become masters of it to some extent, but this is not the case and we have fallen into a trap woven by our own virtue as happened to the poor (foolish and somewhat braggart) Icarus: the fact that we can name Nothing does not summon it into existence in the first place, but even as a mere name it has aroused in us human beings a dark fascination and powerful fear ever since we were able to conceive it. Yet a solution, along with logical errors and emotional turmoil, may be within our grasp. With respect to our question about the relationship between Nothingness and the Universe, rather than struggling to think about how and whether the Universe could be born *out* of Nothingness, it might be better to engage ourselves in thinking about a Universe born *in* Nothingness.

This option has a counterintuitive and, perhaps, even an annoying tone. In fact, on a level that is not carefully pondered but intuitive and instinctive, Nothingness does not appeal to us. Especially as Westerners, it is the name for everything that stands in opposition to what we are, and the gateway to the realm of nothingness seems dangerously similar to death. Here is another protagonist we could not fail to visit on this journey. Death is in fact what makes us psychologically human.

Man is among the known life forms, the one that most intensely manifests the awareness of his own inevitable finitude and that attempts to oppose it in the most ingenious ways. From the most impressive architecture to funerary rituals, exorcising the fear that the idea of death arouses in human beings has been, and still is, a crucial driver for our evolution. What afflicts us, today in a manner probably akin to that which might have troubled one of our prehistoric ancestors, is the uncertainty that opens up beyond this threshold and which none of our efforts seem able to fill.

The ancient gods, although anthropomorphic and subject to the risk of violent death, did not suffer the indignity of time, and in some cases by reflecting the cyclical idea of time, could rise from their ashes, or escape from the captivity with which death, with respect to them, presented itself. That is, they did not suffer it as an inevitable or final condition and could escape it by merit or cunning.

The God of monotheism is completely detached from the material dimension, so from time and consequently from death. Its action is ultra temporal and no end, no term, can apply to what is now called "God" Such a divinity, on a psychological and anthropological level, satisfies more profoundly than other spiritualities, man's need to appease the torment that arises from the meeting of the instinct for survival with the rational capacity to anticipate the future.

The Christian God adds a further element to this dimension: in his becoming man, he breaks the distance that makes the religious message barely palpable and gives it – literally – body. It does not stop there, however. The step-in fact taken by Christianity, and which earned its hegemony over the culture of the time, is the promise to save its faithful from death. The resurrection of bodies on the Day of Judgment and eternal life with the Heavenly Father are new compared to any other religion, and culturally represent an event that in modern terminology could certainly be qualified as disruptive.

Let us dwell on this subject for just a moment longer, which may help us in the following to have a more conscious view of what the God we wish to compare with Nothingness represents in our tradition.

The God of Christians can thwart the eternal strategy of man's power over man, making death no longer an ultimate and radical threat, but a marginal event compared to the eternity of a kingdom beyond life. Faced with the threat of crucifixion, the Christian does not retreat but welcomes it as a travail that will be rewarded. Martyrdom had no precedent in human history, save for the unique exception of Socrates. The father of Western philosophy does not even need God or a projection of eternity to face the end with a serene spirit; he is content with the sense of justice that has been his life-long companion and that he does not want to betray, because the root of his existence dwells in it. The Socratic invention of transcendence has something reminiscent of discovery, but perhaps not within everyone's reach.

But let us go further. As living beings, we are precisely beings, that is, entities, what is seems better to us than what is not.

As entities, we feel we are friends of entities. That is why our culture, whose Greek origins we have explored together, made us imagine a cosmos that in turn, because it welcomes and accommodates us, and because it seems good for us, could not but resemble us to some degree. So it was that, as it was discovered that the voices in men's minds were not whispers of gods but the voice of a soul, the cosmos was also recognised as having a similar essence. The world then began to look at man and his workings. The mark of this living gaze, as it is of all life according to some economists, is "work", that particular way of acting that does not manifest itself on the wings of mere spontaneity or instinct but is oriented towards a goal.

Work is acting in view of a purpose. The very concept of purpose implies that there is a consciousness capable of prefiguring, knowing, and understanding it, and of directing the course of action in a specific direction. Keeping one's course and continually reorienting oneself so as to keep the purpose before one becomes a law, a necessity determined by the consciousness of the goal itself and a watershed for the choices of the man who works according to purpose, i.e. who works.

The fact of looking at things and choosing them on the basis of one's own purposes is the essence of life. And things are what life looks at as tools, useful means to realise its own ends; the action of the living is characterised by its intervention in the world in order to unravel the bonds that make nature a single weave. Man's activity breaks down the world: to make tools, food, to better understand it or to pass on its events.

What are all these things that are abstractable from nature; things which are separable from the whole. The dissolution of this original relationship of the thing with nature, in its disappearance by man, calls into being what it was not. This is the sense in which Plato understands that all work is production, and which consequently makes the "thing" a metastable balance between being and non-being. It is at this moment that the opposition of being to nothingness arises, which interpreted as the negation of everything, becomes Nothingness. It is the desire to enslave the thing by separating it from nothingness that is born with Western thought and that makes Nothingness an enemy of man.

The mutual difference and independence between distinct things that man embraces with its own mind, the not-being-other of each thing exists in view of human planning and his desire to make instruments of them. It is from this "thingness" (a term we also borrow from Heidegger's lexicon) that goals impose their ow needs on action and become horizons of meaning for its consciousness. But this meaning is not univocal, it is plural and evolving, like the purposes of men and the values of the ages, and it is precisely from this provisionality that the discrepancies arise that arouse the dismay of Nothingness is considered the cause and Being (emblematically under the name of God) the cure.

So far, we have attempted to show the network of conceptual relations between Nothingness, Being, the Universe and God, and only marginally have we entered the question on a broad theological level. It should be made clear, before returning to the question of God and his surroundings, that we will not attempt this journey with the tools of theology as this discipline is understood in contemporary times. Today, it is almost exclusively beaten territory in the fideistic sphere and especially within the monotheistic religions; the God of modern theology is, overall, the God of the Bible. We, on the other hand, will attempt a much broader overview and, like all the reflections on our journey, it will be a journey that will take us back to the Greek heart of the matter.

We have already mentioned it: the Greek gods are not the cause of nature but are part of it and – albeit with fewer limitations than we humans – are subject to the laws of nature and fate.

The traditional Greek gods – i.e. before the advent of philosophy – were personifications of natural forces, archetypal feelings and social roles.

They did not answer questions about the origin of the cosmos or the meaning of human existence; rather, their vicissitudes and even their origins (theogonies) were used by the culture of the time as metaphors with moral purposes, as stories useful to exemplify and pass on the values of a civilisation far older than the philosophical rationality with which we, today, identify it even without being fully aware of it.

The gods of Greek mythology, as is also testified by the feeling of great superstition that characterises both Greek and Roman classical religiosity, are useful vademecums for orientation yourself in the concrete world. Unlike the transcendent conception of God, which was certainly born from a syncretism between monotheistic spiritualities and the reflection on the origin of the cosmos from which philosophy began.

The Greek nature/cosmos is conceived as an immutable background that no man and no god made; it is already to some extent endowed with a divine essence by the fact that it is uncreated and eternal, unlike everything else that manifests itself in it, which is instead subject to the law of becoming and therefore destined to end.

We will not deal here in detail with how it came about that, thanks to the confluence of monotheism and Greek rationalism, the gods were replaced by the one God, and this became a constant point of contention for philosophers and even scientists as well as priests.

The idea of classifying the study of the first cause of nature as theology is an innovation we owe to Aristotle. The arché/principle thus first becomes simply *divine*, and over the centuries of translation into translation and even tradition into tradition it finally becomes God.

In this historical process, it acquires a whole series of attributes that have been peculiar to the cosmos, first and foremost eternity and totality (described emblematically in the words of Dante's *Paradiso* in which it is said of God that "uncircumscribed, all things circumscribe"). It is that same conception with which we today, believers or atheists, understand the meaning of the word "God". God thus becomes something akin to the proper name of nature; nature becomes a person, and his name is what is invoked as an antidote to the thought of the end. Can such a God be compatible with a Universe surrounded and even emerging in the bosom of Nothingness? The question sounds irritating, at sometimes scabrous to the point of blasphemy. But that is from the point of view of institutional religion. There is indeed no scandal or heresy in the realm of thought, and it is with those categories of analysis that we will continue this journey.

The god of the philosophers is not the god of superstition, nor is it the god of religion; the latter is the recipient of an emotional need, it is aroused by feelings and embraces expectations and hopes. Reason, on the other hand, orients itself to the thought of the beginning, as well as to that of the end, from another prospective. It investigates the possibility in the light of the facts we have available to us experienced. It attempts to sketch out a structure, a functional scheme for understanding its meaning and implications in relation to other realities that influence us and which we therefore imbue with meaning. But reason dwells in the very mind in which fears, and trembling arise, and the idea that it is possible to disentangle one from the other in a radical manner, as if the soul were a territory that can be partitioned, is an illusion that rationalism often stumbles into. It is precisely this stumbling block that demonstrates the inseparability of the multiple perspectives with which we look at the world and at ourselves; and if we can briefly limit one side or the other by giving preference to the categories that at a given moment are functional to our aims, nevertheless precisely these aims, the goals we set ourselves to achieve, spring from the uniqueness of our human being.

The mind that asks the question "where do I come from?" is never flattened on its calculating nature alone; it is never *just* a mathematician, a cosmologist or a philosopher who wonders about the origin of the world but also the man who dreams and fears, the child who has wandered looking up at the sky and who, beginning to perceive the passing of time and what it inevitably takes away, asks a question of meaning that invests him with his full humanity. Asking for meaning is not a human reason but a man with his reason and his emotions. And the element that connects them is doubt. Both calculating rationality and feeling cannot help but be confronted with uncertainty.

Reason knows that it cannot be universal. It knows that something escapes it, that predictions, however accurate they may be, can never describe reality with absolute exactitude: it knows that any discourse requires a starting point for the discourse, an original postulate, something to which the arguments with which the discourses are constructed can be attached.

On the other shore, the passions – as we usually call the dark side of emotion, wanting to emphasise the influence of them against our will

-drive us to seek a safe harbour for what we are unable to explain or know. Death. The furthest horizons of the past and the future. Meaning is something we seek almost by instinct and which every form of consciousness that has translated into culture has not been able to do without attempting. And we are not satisfied that the answer to our questions are tales or fairy tales, but we want to believe them, we want them to be "true" and more than sometimes we are willing to die to affirm the "truth" of our answers. Reason is aware of its limitations and, in many cases, suffers this awareness and attempts to remedy it. This suffering is no longer reason, but already feeling. Disappointment, failure, are not matters that are understood with the intellect but feelings that are felt in the flesh. The passions do not accept doubt and seek something like truth to dispel it definitively. Like twin stars reason and passions gravitate around a common centre they cannot reach, which constitutes them in their difference and unites them in their intention.

Two paths do not reach the same goal. Religions follow the path that seeks to heal the wound of doubt through faith; that is, through the persuasion that the answers to our questions are well-founded. Faith – of course not only Christian faith – is a gift, an answer to the questions that torment and burden life. To doubt appears as a misfortune, in some cases a mistake. In this way, religions are forms of culture that have evolved by eliminating the legitimacy of doubt and transforming it in many cases into a moral and even civil guilt. Heretic, unbeliever, infidel are just some of the names for the stigma of those who do not give up questioning, testing their own convictions.

1.4. Not only Reason

The history and destinies of the rational path are less linear. The weight of doubt, the fear of Nothingness are all emotions that do not loosen their grip on even the fiercest rationality. Surprisingly, it is precisely when this grip seems to be loosening that reason retraces the path of religion and, abandoning doubt, yields to faith in its own convictions, thus renouncing itself. This is what happens to certain positivism and more generally to what falls under the name of scientism. In fact, a religion of science in which doubt is bracketed, maintained in form but not in substance. It is worth making a few examples of this way of thinking that has much to do with the central theme of our reflection.

A story with tragic and in some ways still very much alive today features the Hungarian doctor Ignàc Semmelweis. The backdrop to the story is the Vienna General Hospital, the most avant-garde hospital of the time and the one where the young Semmelweis was appointed assistant surgeon and obstetrician in 1846. At the time, the mortality rate for women in childbirth was around 10%; a discouraging figure, but not particularly far from the average of the time. However, what caught Ignac's attention was the fact that, under the previous hospital director, this figure was nine times lower.

With a very modern and rigorously scientific approach, Semmelweis considered that such a discrepancy could not be accepted and began to examine the variables between the current and previous management. It emerged that in the time of Dr Boër, the previous director of the hospital, autopsy investigations of post-natal mothers were extremely limited, whereas the current director Klein, whose assistant was Semmelweis, required his pupils to carry out at least ten autopsies every day, which were consequently conducted in between visits to the patients on the ward.

The enigma haunted the young doctor, who was led to the discovery by two further clues: the first was that in the same hospital the mortality rate of women in labour was still five times lower in a delivery ward where doctors did not usually visit and which was run only by midwives; the second was the death of a friend whose autopsy he conducted showed him the same lesions he used to notice in deceased mothers in childbirth. The friend, who had died after a short illness, had injured himself during an autopsy on the very corpse of a patient who had died of puerperal fever. It was then that he speculated that it was precisely the activity of the surgeons that could be the route of contagion. Remember that the season of microbiology had not yet begun, and Pasteur had not yet conducted his revolutionary research.

Semmelweis succeeded in giving substance to his theory by applying an actual experimental protocol involving frequent changes of sheets in the wards and washing hands with a disinfectant solution for surgeons' hands. It was precisely this last initiative that brought about his downfall. Because gentlemen do not wash their hands, and because it was blasphemy to suggest that doctors could be *anointers*. Science at that time thus revealed its darker side, becoming a practical liturgy far more profoundly than an exercise in free thought. The ostracism caused him depression, some mental hospitalisation and a social and professional stigma from which he never fully recovered, even though his return to the profession gave him more opportunities to prove the validity of his intuition and save thousands of young lives as a result. By a twist of fate he died of that very fever, which we now call septicaemia, contracted from a wound during an operation.

Among the great names in science, even among physicists, more than one has suffered their discoveries to the point of sometimes repudiating them. Among these is Max Planck, to whom we owe the solution to the enigma of the black body. Without going into the more technical details of the question, which is fascinating as few others but not necessary for our purposes, the German scientist wanted to propose a hypothesis for a solution to a problem that had been raised forty years earlier and which consisted in finding a relationship between the temperature of an ideal physical object and the energy it absorbs and consequently reflects. Planck's idea was to overturn the classical dogma of the continuity of nature and to assume that energy could only be transferred via "packets" of energy. Experimental data confirmed the physicist's mathematical hypothesis, and even today, the minimum units of physical quantities, known as Planck units h, are dedicated to him.

The success, however, of his idea was not particularly pleasing to him; it was too far removed from his sensibilities and radically disrupted his world view. His reason, playing with nature, had shown him a spectacle that was not within his grasp to accept. How could it be true that nature was not continuous? How could it be possible that space cannot be divided infinitely, or that there is an ultimate time scale, as if God had notches on his watch just like on ours? It wasn't the solidity of his thesis that made him doubtful, but rather the very conclusions he had been able to prove. The world "really" seemed too far removed from the vision he had become accustomed to and relied on all his life. And indeed, quantum physics, which takes its name from Planck's quanta, has entangled the current framework of scientific paradigms more than any other discovery.

Einstein is also among those personalities who have not always had a good relationship with the extreme consequences of their theories. It is still in the context of the emerging quantum physics that the most famous scientist of the last century struggles with the idea of wave-matter dualism. How can matter be both one and the other? How can it change nature, or essence to use a term even more loaded with metaphysical implications, by virtue of its observer's gaze? As a realist, Einstein expected more than that: he wanted the world to be given, a fact to be discovered and not a form that changed under the eyes of his investigator. How else could one be certain of the solutions to the great puzzle of the cosmos. In the famous quotation from the correspondence with Born in which he says that he is incredulous of the possibility of God playing dice with the world, all the emotional tension emerges, all the range of convictions and emotions that cradle even the most celebrated rational mind of our age.

There is, however, a substantial difference between the incredulous, even bitter attitude of Planck and Einstein, who remained somewhat *faithful* to their own worldviews, yet did not reject the "in-credible" news of their reason. In these cases, science has not succumbed to scientism; it has not been tainted by that superstitious arrogance that renounces all doubt in order not to waver.

The distinction is not between science or religion; both can indulge in the illusory security of the beaten path, preferring a fallacious but familiar path to the weight and challenge of doubt. Both can succumb to the fear of the unknown and react to it, like any cornered animal, fighting to the bitter end to save themselves. However, there is no external enemy to defend ourselves against, for doubt dwells within us; uncertainty is an inevitable doorway of consciousness. More than a thousand years ahead of Cartesio and his *cogito ergo sum*, St Augustine had already recognised the inseparable link between man's being and doubt.

When we look at the Universe and the Nothingness that frames it, or at least that frames our fears in turning to nature, our contemporary gaze is surprisingly close to that of any of our ancestors in the centuries and millennia before us.

What is astonishing and astonishing is that what we know about the cosmos is not qualitatively so much better than what we could have discussed with a Babylonian or Aztec astronomer. Let me explain: what we have not made the enormous progress on that we often tend to take for granted are the "big questions". While we have in fact discovered myriads of sophisticated and incredible mechanisms on which the phenomena we can observe depend, and even spy on today, we are, on the other hand, more or less at the same distance from a "solution" to the question of "why do we exist?", or "what is Nothingness?", than the Neanderthals who painted the Lascaux cave. This is because science, or rather, let us say it in full, *knowledge*, is not a collection of facts, but a process. That extraordinary and somewhat frightening path we can only travel by embracing what we do not know.

Knowledge needs Nothingness in order to be; it needs us to have an inner emptiness, a sense of bewilderment pressing enough to make us devote time, effort, energy and resources to filling that void. From this comes frustration as soon as we realise that filling it is impossible, and that every answer opens up more questions than it was able to resolve. The Nothingness we fear so much is, looking at it with enough detachment not to be overwhelmed by it, the engine that offers us some of the most peculiar ways of being human.

What we have called the *big questions* are those whose answers can radically change the way we live; those from which our daily behaviour and actions can receive a decisive boost.

Keeping anchored to the theme of Nothingness that is cogent for us here, "what will become of me after death?", which in practice is like asking whether or not I will have to face nothingness and whether or not I have any chance from this confrontation to still come out in me; will there be a day when I will be nothing? Well, the answer to this question, which can in no way come from science as a field of knowledge based on empirical facts, can be a watershed for us in everyday life. In this sense, Nothingness is *by no means a trifle* – pardon the pun, but in this case it is very serious – because it is solid, it is a potential cause of our acting and feeling; Nothingness in the moment that we are able to conceive it, at the limit even to misunderstand it, nevertheless moves us like gravity, hunger or love.

1.5. Doubt in front of the mirror

A strange position is occupied by another type of big questions, which resemble the ones above, but may differ from them in some ways. For example: "What are all things made of?", "How did the Universe come into being?". These, too, are questions which the answer could radically change the way we see and thus also relate to the world; likely, our behaviour would not be the same if God were a fact like bacteria or weather phenomena. Today we are more inclined to ignore the voice of a God we do not see or know, much less to eat without washing our hands or to leave our umbrella at home if the weather forecasts rain.

If God as creator of the world were to become a fact, the lives of all of us would change, and not by a little because it would be much more difficult to ignore Him. Similarly, and with all due respect to Ivan Karamazov, if we had the certainty that God does not exist – impossible to have this certainty as well as the opposite since it is not a fact but a perspective that is in principle neither verifiable nor falsifiable through experience – many moral constraints, many taboos would probably be forgotten.

Now, if God as the cause of nature is not among the possible phenomena that can be shown because of its transcendent nature, it is not in principle implausible that someone at some future time in some part of the Cosmos might understand the mechanism that started what is our Universe and that, who knows, it might reproduce that phenomenon in some way having understood the relationship between matter, energy and information that is needed for an event like the Big Bang to happen. Or again, a question as old as the question itself and on which we have already asked ourselves concerns the extension of the Universe, the parts of which we wonder about and, almost like explorers of new lands, the boundaries. Again, if we got a list of objects, planets, molecules, stars and black holes as an answer and were certain that *this was the whole Universe*, we would probably not be satisfied. That is not the kind of answer we are looking for; that is not the feeling we imagine associated with the big questions. The thirst for knowledge that flows from them does not seek "things" but epiphanies. Reason alone cannot be satisfied by knowledge to quench our agitation and that sense of emptiness that makes us move, our passions must also take their toll, and this must be something that can also be felt by that part of the mind that remains uneducated no matter how hard we try to tame it.

The void left by the sense of Nothingness cannot be filled by a simple accounting exercise. Similarly, our scientist friend who, in a few eras time, could reproduce a Big Bang, would probably not stop wondering (as we have already said) who or what preceded this "big explosion".

There are two watersheds that divide the great questions of science from the great "existential" questions, and which ultimately distinguish those which are questions in the proper sense from those questions to which the answer can only come from us.

The first is to admit an answer. However far removed from our current understanding of the world, even if we could never conceive of it, the Universe – just to stay with our theme – has happened in some way; it is a phenomenon whose aetiology can in principle be described. That is to say, there is one correct answer, one description of the facts that corresponds to them in contrast to all the others. By asking instead what the meaning of life is, or why something exists instead of Nothingness, we are asking ourselves something that perhaps does not offer an answer, or – in the case of Nothingness – offers only an "indirect" answer. The meaning of life is not in fact a quantifiable datum, and each person will have to confront himself in order to try to construct his own path towards that answer; on this level, Nothingness is also not a thing, and therefore cannot be compared with an object or a phenomenon by finding similarities or differences between them; since Nothingness is a viewpoint on the world, or rather that which systematically escapes any view of the world is not comparable with the other things we can embrace with our senses.

The second – formalised between the 19th and 20th centuries in the field of epistemology – is that a scientific question is only such if it admits *in principle* the possibility of being falsified or verified (the two points of view are distinct and full of implications, but do not concern us specifically); that is, if it is possible to hypothesise a circumstance, event or phenomenon that, if it were to manifest itself, would imply that the question would have to be answered negatively or affirmatively, and this is the case with the fate of consciousness (or soul, if you prefer) after death. That question of what will become of *us* is legitimate, perhaps even one of the signs that sketch the essence of human existence, but we cannot expect science to give us a response. There is, at least at present, no possibility of something like consciousness manifesting itself in the quantifiable ways that a scientific phenomenon can.

We are consciously shuffling the cards a little and overlapping reason and science, and in a moment, we will also add a few thoughts on philosophy. What legitimises this assimilation of the three perspectives, which when analysed in detail are points of view and ways of thinking with profound and significant distinctions between them, is that, at least in their foundation, they are united by a similar way of proceeding. Faced with a question, they test possible answers. By asking "what exists?" we could compile a list in which we include everything that meets our criteria of qualifying something as existing.

Then the book you are reading, and yourself, and your socks, and so on. Could God and Nothingness fit into this list? Perhaps yes, perhaps no. They certainly would not fit in as an effect of applying the same categories that made us add ourselves and the book to the list we are reading; and again, could both fit in? We will return to this point later. For now, let us be content to consider, and hypothesise without fear of being proved wrong, that instinctively the way in which we regard the objects around us as existing is not the same as the way in which we regard God or Nothingness, and probably not even the universe, as existing. And this without needing to refer to Aristotle's categories or any other profound and careful logical clarification of our language; our instincts, sensibilities, or common sense, regardless of personal convictions and faith choices, do not conceive of God on the same level as things that are in space and time. For we have been prompted throughout history to ask ourselves what they are made of, and what kind of connection they might have with that special thing that is ourselves. We have also been able to find an answer to this question by discovering the chemical elements, and subsequently the atoms, and within them the subatomic particles.

Today, we can actually answer the question "what is the Universe?" to an exceptional degree of approximation by saying that it is a collection of a few elementary particles – quarks, electrons, gluons, etc. – but we would still have missed the point. The Nothing would still remain just behind our horizon, and we would feel its looming over us no less than children frightened of the dark. We are in fact looking for something deeper, for a kind of truth that appears on a different and further plane than the things it refers to and says exist or do not exist. No list, no set or collection satisfies this requirement of what is beyond; we could even use the term transcendence to refer to the form of an answer capable of satisfying this desire for completeness, totality, universality. Precisely because these attributes are not things, but modes of being common to several things, entities and experiences such as to subsume and exceed them.

What we seek, more or less consciously, are recurrent forms, patterns, to put it in a language more akin to the contemporary one, capable of organising our knowledge and revealing links and correlations between elements that we previously thought disjointed. You and me, one of the swallows circling outside the window as I write, even the wooden window frame, we are made up of more or less the same parts and in proportions not so different from what one might think. Perhaps it is also for this reason that our pre-understanding of reality leads us not to stop at the parts but to investigate the relationships between them, judging that the truth, the epiphany we seek to quench our thirst for knowledge and with it our doubts, has more to do with relationships and proportions than with the parts themselves.

Deepening our knowledge about the world, it seems that our rational instincts were right and that among the many biases that make us fall prey to deception and illusions, at least when we look at reality and investigate the deep structure of matter, our pareidolia has been confirmed. In fact, if we go beyond the elements of the periodic table and go to the level of the particles that make up atoms, we discover that in the 100 grams of pasta that we ate for lunch there are the same quantities of quarks and electrons that are contained in 100 grams of clouds or the moon. What changes is not the bricks the world is made of but their arrangement. These bonds allow us to simplify the complexity of the world, to see it as composed of a prism with many faces instead of as a chaotic jumble of many distinct things. Modern physics today calls the last frontier of theological research the *theory of everything,* a very poetic name that reflects with great sincerity the demand that has been driving the research since long before mankind did science and philosophy in a fully conscious manner.

And here we return, by another route, to one of the points that have been a base camp of our journey. The search for the whole, for a universal explanation that leaves no grey areas to undermine our gaze. Man's dream is to embrace, at least with his mind, everything. More: to go beyond things and master their essences to overcome even their transience. Inevitably, if we are the protagonists of this epic and imagine ourselves the heroes, the role of antagonist could only fall to Nothingness. This strange duel, which is more or less the driving force behind every form of culture and civilisation that has ever appeared on our planet at some point in our history, has nevertheless revealed itself, in the eyes of a small nucleus of humanity overlooking the Mediterranean, to be a game with neither winners nor losers; a dance rather than a duel. And taking part in it is something so profound that it contributes to defining the very essence of humanity. The world's becoming is the soundtrack, and the rule for remaining in harmony with it is to accept that you cannot stop it. To take part in it to the limit even as protagonists if you manage to take centre stage for a moment, but not the directors. That role remains behind the scenes of reality, unavailable to be taken on by us who are on this side. Taking part in this game that is the soul of humanity means accepting Nothingness and renouncing the illusion of removing it from our horizon. Every discourse, research and thought that arises from this awareness is philosophia. Philosophy is also every authentic science and every religion. Science that accepts to contribute to humanity's journey towards knowledge, aware that it will be taking a step that is by no means definitive and yet nonetheless significant. Which knows itself to be temporary, as any acquisition is, and which does not seek to make Nothing vanish but is content to move it a little further.

We are unable to say whether it was Nothingness that aroused philosophizing or whether something like philosophical thought was always crouching in the human soul waiting to recognise something that would arouse it. Rather, however, we are firmly convinced that between Nothingness and philosophy there is a foundational, insuperable, and constitutive relationship to our way of being in the world. This is why we want to try to reason together about what this word means, which denotes a way of thinking, being and relating to the world that is superficially taken almost for granted but which, like the Nothing that complements it, can escape the grasp of an improvised definition.

1.6. *Thinking for real*

Philosophy may at first appear as the attempt, particularly by Plato, whose thought we have already encountered, to provide an ultimate and definitive answer to reality; to discover the absolute and proclaim an incontrovertible truth. To vanquish Nothingness once and for all by overcoming faiths, beliefs and hopes with necessity: the impossibility that this that is different from how it is. That law, or pattern, or pattern that goes beyond the singularity of things and grasps at once their actuality and possibilities, exhausting and capturing the totality of the thing. In this desire, which at times has elevated itself to a claim – because not even philosophy is immune to the illusions our minds are capable of weaving – the intrinsic unattainability of its goal is revealed. Because if – perhaps – of the thing, or at least of some things, it is conceivable that one can grasp the totality of it, the same is not reasonable to wish for the totality of things.

Philosophy, in short, was born in the desire for the whole and is, well before modern science, a great collection of theories of the whole that are nevertheless known to be incomplete. It was not science that showed this vulnerability of philosophy; if anything, on the contrary, it was philosophy that taught science, which sometimes forgets this, that there is no definitive step and that everything is surmountable and destined to be revised, corrected, inserted, and subsumed within new ways of seeing the world. Not even religions have made rational thought and its dreams of the absolute into conscious illusions. Philosophy, on the other hand, has taken seriously the entities that populate the world of faith and some – like God – have even lent it to it without ever ceasing to question it and to confront the implications of its being.

However, science and religion, from opposite sides, but which in the radicalism of certain fanatical positions such as scientism and fundamentalism have sometimes come to resemble each other in excluding all other points of view, have shown themselves capable of living without philosophy. That is, of living in a world of illusions that ignore their fictitious nature. There are two ways of living without philosophy: that of those who ignore the facts and replace them with their own narration of the facts and who, like Ptolemy's cycles and epicycles, invent an excuse for every derogation that shows itself to their interpretations and predictions. And that, much more subtle, of those who reject all incontrovertible truth. At first glance, this also appears to be a philosophical attitude and, in some cases, what we nowadays confuse with relativism or perspectivism, an authentic way of thinking philosophically and rationally. But in order not to fall into contradiction, the denial of any definitive truth cannot constitute truth in itself; this would be the case of scepticism based on a paradox that has been known for over two millennia. What can happen is to live *as if* nothing is definitive. Abandoning all pretence and search for coherence and making the fluidity of will and the opportunities that come to power a fluid script to be interpreted without ever looking back or ahead. This way of life – which is in fact an ethic that is no longer known as such, since if it were known it would become a choice and would rest on a guideline, and the latter would re-emerge as a truth that attempts to deny all truth – is pre-philosophical and in fact pre-human. It is the fair way of life. Free of any project, immune to the weight and the law of purpose that serves as the measure of action which, without a goal, can no longer be said to be neither good nor pernicious, neither just nor unjust, the agent almost loses his nature as a subject, he becomes confused with his action and his individuality fades into the nature that is the theatre of his action but of which, now no longer a subject, one cannot even be sure that he is aware. As a reward for the nothingness he becomes, he has defeated the Nothing, he no longer sees it; he does not feel its weight or even the disquiet of its impending presence. But without Nothingness, without the awareness of finitude, is one still human?

Philosophy, before and more intensely than it is a way of thinking, is a way of living. It is the ethics that characterises anyone who acts considering his or her own finitude and that of the world in which his or her actions unfold. The fact that it is oriented towards unveiling the *Truth of* the world does not contradict the insight that this transcendent and all-encompassing truth is elusive. What follows from the philosophical life is that the world has *meaning*, that it is accompanied by *values*.

Philosophy is the way in which we live conscious of the opposition between Being and Nothingness; consequently, an image emerges of the world as a theatre of the rising and fading of things, characterised by a temporary *being* and whose accidentality is a capital question in the eyes of any accomplished consciousness (or more precisely, self-consciousness). A further reason why Nothingness can appear to us as an enemy thus becomes apparent: entities and phenomena that were once not – or at least not within a horizon that is reachable from our perspective – suddenly happen. They often literally fall into our lap.

They come into our lives, disrupting the course of events that we had preconceived and diverting us from our goals. It is quite evident that they can appear to us as evil. The Nothing seems to be the crucible from which tumults emerge, clashing with what is already there and sometimes usurping its place, driving things back into the same Nothing from which they had emerged. That a similar fate befalls us, and our projects frightens and terrifies us, making us feel all the impermanence that we try to defeat and remove as inscribed insuperably in our flesh. But these are not the only realities that emerge from Nothingness. There are also favourable events that are no less inadequate than catastrophes; fate can be both favourable and unfavourable, and if we are more attentive and better remember the eventualities that have stood in our way, this is because the evolution of our species has been favoured by a good memory for what is dangerous so that we can try to find a solution, while remembering that we have had good luck improves the life of the individual but does not bring great benefit to the community.

Yet, the joy of good fortune could not fail to leave its mark on thought and civilisation, and if we have baptised the source of misfortune as Nothing, we have given the cause of good fortune the many names of God (and before gods).

Nothingness and God once again display certain traits that are in the opposes, yet they have them in common. Their radical distance from us and our world; what can be called transcendence, and which is one of the ways in which the absolute is manifested, can be recognised in one and the other. This analogy has sometimes produced cultures and spiritualities that have elevated both light and darkness to divinity; at the border between East and West, one example was Manichaeism. However, it has not infrequently been perceived as scandalous, or even threatening.

In our constant search for fixed points, the opposites, the metaphysical good and evil, it is good – precisely – that they remain distinct and recognisable and that at least on those we do not have too many doubts to confuse our three consciences. So while the solutions to the problem given in the East are more fatalistic and have accepted Nothingness as a given and in some cases even as a possible solution to pain instead of its cause, rationalism in the West has thought of inventing an order and a language that would avoid any risk of confusion: God is Being (or the source of Being) and Nothingness is the opposite. It is a pity that language works better to indicate things than to solve problems that arise from language itself. The issue in fact, once "settled" is not one we are very happy to talk about. We assume it, but we do not reason about it willingly. Philosophy has thus created an additional problem in its attempt to solve one that still persists. She too then, true to her mission of unravelling the mysteries of the world and tracing the boundaries within which Nothingness and Being (whatever its name may be) contend, stands in defence of the threat she herself has evoked. That is why she has taken the question of God tremendously seriously.

The first cause of nature, divine or otherwise, in order to be able to effectively oppose Nothingness, should be able to limit its freedom, not to leave it free, but to harness it in a predictable or at least conceivable plot. It must, in a certain sense, be able to pass through the gates of the Nothing and legislate even in its realm. And so, we often imagine the laws of nature to be not only binding descriptions of what is in the world, crossing the boundary of Nothingness and entering the realm of Being, but also constraints on what is not yet. The Nothingness in this world comes to lose its absoluteness; it no longer transcends the world and Being, but as it awaits it.

Even unactual entities seem somehow only to rest in Nothingness and await the call of Being that will summon them according to its own laws. Their appearance on the scene of the world is only an apparent novelty, they are in fact already there by virtue of the fact that they are included in the possibility of Being and foreseen by necessity. The impossibility that they are not annihilates Nothingness. Since Nothingness speaks the language of Being and respects its law, it vanishes, and everything simply *is*. Within this perspective, Nothingness loses its capital letter and no longer stands on the same level as Being. It is still a bogeyman for human finitude, still events can certainly happen that mess up our plans; but it is our problem, an accidental limitation, no longer a matter of principle. Nothingness (minuscule) remains as our problem but no longer as an insuperable structure of nature. Then the imponderable simply becomes the imponderable, and sometimes we cause ourselves to call it an *error*, so that we take on its imponderability as if it were our fault.

By now we should be often accustomed to the vicious circles, and then evil becomes a consequence of our being, a reflection of ourselves looking at the world and doing evil because we are unable to listen or obey the call of Being, in these cases often explicitly called God. It is then impossible to understand the reasons for certain atheism, because such a God can turn out to be a judge who punishes without distinguishing between guilt and malice.

Partial view of the theological question, but understandable when it arises in the mind of one who has cogencies and urgencies that do not give him time and opportunity to spend his life in meditation. And the vulgate that makes evil the sign of guilt poses (and has done so for centuries) the dramatic question of innocent evil.

Man then in his philosophical living does not give up trying to find solutions, and so if Nothingness is no longer absolute then it can also be conquered by me; if it is an accident, it is within my grasp and the way to get to grips with it could be knowledge. Thus, we return to the search for truth. This time, however, the risk is that of scientism and of making truth that immanent substitute for the transcendent God that does not harmonise with our good intentions when they fail. Uncertainties sound like nefariousness because, if evil ends up coinciding with ignorance, they echo with guilt and science can turn into a new religion not devoid of fanaticism. But philosophy is the opposite of fanaticism. Both antidote and poison it seeks the truth of becoming, but it is either truth or becoming. Philosophy knows this but does not renounce it. Then, just as animality is to live the absence of purpose in order to remove the possibility of something like Nothingness opposing it and tormenting us, philosophy is the life that embraces doubt but does not bend to it; it is the life that is born inflicting the hardest blow on itself, accepts the split but does not submit to it even though it knows it is insurmountable.

Life is the only possibility of facing the contradiction that is in-eradicable on the plane of logic and thought. Philosophy is the life that makes knowledge an ethics, that makes knowledge act and suffer.

When knowledge finally transforms Nothingness by making it a void that life tries to fill with all the limitations we have already discussed, so the small Nothingness of the difference between points of view can nourish integration, exchange and form the basis of dialogue by nourishing different cultures and stimulating them to evolve in reciprocity much more than would happen in isolation.

If physical space takes on meaning for us in interaction and relationship, so social and cultural space – which are ultimately names for a way of organising our experience and our way of relating to a particular classification of external reality – are fields of possible ways of certain forms of interaction. It is by virtue of the relations that can be manifested in them that these *are*. Relations are the building blocks with which the world seems to be composed. In this way, knowledge is not configured as a collection of monolithic, separate, individual entities; nor is it a single, static thing. The image that most truthfully describes it is that of the web that knots some moments or places more tightly than others, that leaves some ties untangled and that still others hide from view.

Once again, the idea of *what* seems to be an old-fashioned and inaccurate simplification for describing Nature, very much akin to our immediate perceptions reorganised based on practical urgencies – such as eating and defending oneself – rather than a contemplative purpose oriented towards creating truthful representations of reality.

In the aforementioned contexts, Nothingness is that which remains beyond the possibility of relation and specifically of knowledge, Nothingness (minuscule) is that which we do not know and which seems unconnected to the web of relation of which in one way or another we feel we can wield a bandana.

Finally, social or cultural nothingness is otherness, that with respect to which we do not feel any similarity to what we are and feel: the absent applied to a difference on which we do not grasp the contingency. On the opposite side, God-Being is the name of the total network, the complete and enclosed relation with respect to which *nothing* is extraneous. Once again, Being and Nothingness refer to each other when one attempts to capture them through the logic of definition.

Any distinction that claims to be absolute is increasingly shown to be an arbitrary exercise, sometimes useful for particular purposes, but not if the intention is to represent reality.

Nature and culture "function" in very similar ways. We are a part of Nature, and our specificity, a way of recognising ourselves within the network of phenomena we call natural, is our possibility of creating an image of that very nature of which we are a part. Nature is precisely this image *for us,* and the idea of an objective foundation is not part of it as such but is the result of a desire as we conceive this image. It is not Nature that generates the idea of an objective foundation but our will that accompanies the vision of this image. We thus manage – perhaps – to propose a possible closure at least for this part of the circularity that we have approached from the side of knowledge: Nature shows itself as a cultural phenomenon and it manifests within nature. This is not a relativist point of view in the pejorative sense that the term often assumes in debate when one wants to make someone appear as a disbeliever destroyer of values; quite the opposite. This viewpoint, scientifically rigorous and allied with the latest findings in physics, epistemology and cosmology, respects our condition. It does not ape an omniscience that really does not belong to us but expresses a view of the world that is consistent with what we are and what we know we are, and not with what we are not and what – who knows why – we would like to be. It says what of the world, Being and Nothingness we can hear instead of shouting at them as we would like them to be.

We ourselves together with our consciousness are a point of view; with more than others – at least in appearance – the ability to include ourselves among the phenomena that fall within our horizon. A recurring and relatively stable pattern, a pattern capable of including itself in the search for other regularities around us. We are "a look from nowhere". What makes it so special, what to some extent grounds the value of this gaze of ours is the awareness of its finitude.

Once again, Nothingness, even that which we grasp with respect to ourselves, is not only an evil and if it can certainly cause fear, it is also a founding component of value. The fact that this gaze knows that it will not remain perpetually illuminated makes the moments when it is more intense. We are inclined to think that this finitude is an attribute of ours and that Nature knows no end, but this too is a story we tell ourselves rather than an awareness based on experience. For Nature's destiny is something that eludes us. In this, Nothingness acts as a frame and perhaps also as a cradle, since it preserves and brings out what, after all, overcoming our fears and instinctive resistance, is also its value.

Yet, something can be said. Science, particularly physics, offers us many elements on which to reason and, at least attempt, to find some perspective on the behaviour of Nature.

We have already described some characteristics of the cosmos. In particular, we have seen how the expansion of the Universe takes place at an increasing rate. As we come to our conclusions, having tackled the topics we have proposed from various points of view in order to offer as broad a picture as possible, we would also like to put forward a few hypotheses. Let us then think of the Universe as a sphere; to the three dimensions that we in our habit regard as spatial, we add the fourth temporal co-ordinate. Over time, this sphere, like a bubble, inflates and as it inflates – we have already seen this – new space is born. As long as the distance between the boundaries is such that it can be reached by a photon from the opposite side, our bubble can be conceived as *one*; as a Universe. But when the speed of expansion, which as we have seen is increasing according to current measurements, exceeds that of light and therefore makes it impossible for a photon from one side to reach the opposite side, can we still say that we are in the presence of a Universe? If the extremes of our bubble can never come in contact are they still part of the same Universe?

Now, assuming for the sake of convenience that the bubble is still one if been looked from the outside, as from the point of view of that absolute on which we have spent so much time reflecting in these pages, we would see that from two opposite points, two distinct photons each bounce within a new bubble extended enough to be crossed from border to border. Not that the first Universe contains others, but it has *become* others; it has ceased to be what it was, the network of relations that constituted it has expanded to the point that it can no longer embrace all the phenomena that manoeuvre within it. That network is no more. But other networks can be identified and encompass the point of view that we are.

As far as we know today, light is the boundary that circumscribes the possibility of relationship; what light brings into contact, the phenomena it can illuminate, is one universe. But if distinct phenomena can be illuminated but can never fall within a common horizon, do they not behave as two universes? And what separates them, their mutual incommunicability, is something like Nothingness. Not a fact but a relative condition; a way of being of events, their not being (or no longer being) connected. The boundary as the bubble expands, does not cease to exist, but shifts. If we could watch the process from outside, we would see it as dividing into other bubbles. The expansion of the Universe produces other universes, creates nothingness as it creates worlds and creates worlds because it creates nothingness between them. We set out to answer whether such a reality, which continually creates and produces Nothingness, is reconcilable with the thought of God. Well, perhaps the only God who is conceivable beyond the confines of a fairy tale is the one who is not only compatible with Nothingness but coexists with it. Without Nothingness, God is a despot. Nothingness is also the condition of possibility of freedom, the spark of meaning and value; and could the divine ever be conceivable without meaning or value?

This is only an image of course, an illusion. The fruit of a description that attempts to embrace with our categories and subsume with our habits something that radically exceeds and exceeds them. This, or something similar, is what our mathematics sketches when it lingers beyond the boundaries of the world and tries to legislate in the realm of Nothingness.

A similar picture, however, can also be painted by pushing the boundaries of another Nothingness: the event horizon of a black hole.

Having crossed that boundary, the light is trapped, the Universe in which it has ignited is closed to it forever (whatever that means). Gravity makes space "steep" and causes light to slide further and further towards its centre, that which we call the singularity. The pace of this fall is creeping, as is that of the expansion of space at the boundary of "our Universe". What falls into it remains, at every instant, bound to all that is swallowed up in the same instant. Precluded from everything else and to which the rest remains obscure. As in a bubble...

Conclusion

To play along, among those of us who are writing to you as well as with you who are reading us, now at the end of this journey together, we can apply the same philosophical approach so far to the demonstration that Prof. Cantalupi gives in the following pages.

Without needing to return to the detail that can be found on p. 173, the basic idea of the demonstration shows that it is possible and mathematically possible for 0, understood as the mathematical presentation of Nothingness, to give rise to 1. And this is an exciting fact and – at least for us it was – such as to arouse a sense of happiness like that offered by certain poems or panoramas. But if we want to mix mathematics with life and attempt, we repeat, as a game, to exercise ourselves in a contemplative activity in the manner we have endeavoured to make sense of above: are we sure that it was all Nothing/0? Without us, could it have entered into relationship with itself and generated that 1 which appears at the conclusion of the demonstration? Or is it from us and our being, however accidental and finite, that it could have borrowed that bit of Being with which to generate the 1?

The questions we have just asked, although formulated almost in jest, lead us to profound reflections, and indirectly from them, unexpected answers unfold. One of these concerns the temporal location of the entities of the cosmos. A serious theory on the birth of the universe cannot foresee that an entity that should be primal can move from something that already exists. The "Being-us" or (we shall see later) the quantum vacuum are something tangible and cannot be without an entity that comes before. A complete theory of the genesis of the universe, on the other hand, must start from the total absence of "something" that is present before "Being-us" or the vacuum; in short, it must start from Nothingness.

BIG BANG, WHITE HOLES AND THE UNIVERSE FROM NOWHERE

A few months passed between the first edition of the book you are holding in your hands and this re-edition and translation. Fans of science and astrophysics in particular may have come across Carlo Rovelli's latest work: *White Holes*. If you have come this far, you share a fascination for borderline territories, and with this in mind, the following reflection is born. It is an opportunity to look at new perspectives – which experimentation will perhaps disprove – but which are currently the subject of first-rate publications and which are deserving of in-depth study by professionals throughout the scientific community.

What we will attempt to do is not yet fully physical, but its embryo. The Greek language that has accompanied us from the first pages and that offers itself as a source of technical vocabulary to all languages, including modern ones, would call it *theoria*, which not for nothing sounds more or less the same in Italian, English and elsewhere. It literally means: to look with the eyes of the mind; we can only use those to move through the territories in which we wish to venture.

The question, which we have discussed so far, but without really attempting to answer on the merits, is whether it is possible, and before that conceivable, a universe that comes into being from nothing. We have made it clear that the answer cannot fall within the confines of the natural sciences, that the origin exceeds (transcends) the originate to which we belong as physical entities. However, this does not prevent us from approximating our metaphors and myths to the conceptual constructs with which science describes (thus understands) the world. The origin of the universe from nothingness is a thought that appears almost contradictory to us on first immediate impression. Nothingness seems to us incompatible with a generative act. The old-fashioned thought that "nothing comes from nothing" seems more intuitive to us.

In order to challenge this intuition, which we perceive to be somewhere between a principle of nature and a cultural assumption, we need an ally, as if strength or courage were not enough: science. Or rather: its language, its metaphors, its conceptual constructs. Precise, more authoritative than our dreams with a metaphysical flavour. In response, knowing that we cannot obtain anything like the truth of knowledge, we at least hope to keep within the boundaries of validity. This is not enough for us. The dream of every physical theory is to be able to inspire an experiment that either corroborates or falsifies it; arduous as pointed out in the previous pages is to hypothesise an experiment that proves where and how the universe came into being. But thought can hope to get closer to the truth by constructing more coherent metaphors, more pragmatically effective as descriptions and explanations of natural phenomena. For this to happen, these bridges between thought and nature must be constructed in a manner, in principle, subject to empirical judgement. The only way to do this is to make use in the construction of the key concepts of the scientific paradigm within which one moves. Moreover, to include its axioms in our model as further constraints on speculation, which in this way consistently becomes a special ontology (or even formal ontology).

The origin of the universe, in science, is very close to being something like the Big Bang. And if there was "nothingness" before, how is it describable in the terms of our physics? Can the Big Bang perhaps be thought of as a white hole?

No human eye can – and perhaps never will – get where we are going. No experimental physics seems in principle imaginable that could satisfy our desire to know. But this has never inhibited the adventure of science. Leave the baton in the hands of those who will pick it up. One question. One step.

Our compasses and maps will in short be the pillars of the Standard Model, particularly the theory of general relativity, of which white holes are a deduction. The aim is to hypothesise whether this universe of ours, its becoming and in particular its origin – which we have already mentioned cannot be included among the knowable phenomena as it is intrinsically metaphysical and transcendent with respect to the universe itself and therefore to our very existence – is not perhaps conceivable thanks to this new object-concept that physics offers us humans to feel more at home among the stars and our doubts.

Let us begin.

What is a white hole? In a nutshell, it is a phenomenon symmetrical to a black hole. The latter originated, almost as an invention, within Einstein's theory of general relativity: a phenomenon predicted by his equations long before it could even be thought of as an observation. A "hole" because the gravity manifested by certain concentrations of mass is such that it does not allow even light to escape its attraction; hence "black", because not even light comes out of it. Something that generates "almost a universe" within it, which remains by gravity separate from the rest of the cosmos, or so it has seemed to us until now.

A handful of decades have passed since those early writings, lacking a solid experimental and observational foundation. The consistency of the theory – i.e. the fact that its mathematics held up, if we want to simplify matters a little – prompted the entire scientific community to look to the sky once again in search of objects that were invisible by definition. The traces of these gravitational monsters were of the same kind as the observations that were the empirical confirmation of Einstein's theory: the so-called Eddington experiment, which – after an unsuccessful first attempt in 1918 – was verified on 29 May 1919 by Andrew Claude de la Cherois and Charles Rundle. Mankind had recognised starlight "bent" by the Sun's gravity; now we only had to look for other strange distortions to bring us closer to these invisible objects born of Einstein's thought and described in mathematical language. We emphasise this again: from thought, for nothing had been seen or measured, but only thought of by his mind; perhaps the most famous physicist in the history of science had discerned from the theories of those before him how recurring forms he had attempted to give substance to. Many distinguished scientists still did not really believe in the existence of black holes when, on 10 April 2019, the photograph taken by the Event Horizon Telescope of the black hole Messier 87 was made public.

To return to the question we started with: a white hole – like a black hole at least until the fateful 2019 – is a phenomenon that is compatible with the theoretical framework we call: *general relativity*. That is to say, something that in order to be described does not contradict any of the theory's assumptions (we speak of logical consistency); that moreover does not imply any new assumptions (it is *parsimonious* since it no longer requires axioms); it does not contradict direct observations and is not "disproved" by empirical verifications with which it is supposed to be confronted (that is to say, it is testable or falsifiable in principle).

The mathematics of white holes, at least until today when some further verifications are still in progress, also reveals something that borders on the paradoxical, in fact a white hole and a black hole, seen from the outside, appear to us in the same way. They show a horizon reached at which matter, after progressively slowing down, comes to a halt in an apparently eternal stasis. We call this threshold the *event horizon* because it is a condition in which the curvature of space (spacetime as we will be reminded a little later) becomes "steep" to the point of preventing light from escaping. One might be tempted to imagine this horizon as a place where photons are immobile, almost as if they were behaving like the jib marking the midpoint in a tug-of-war between the gravity of the centre of mass and the energy of the photons; and – having fallen into this temptation more than once myself – we would be wrong. In fact, the time dilation effect, that slowing down to stasis, is only a perspective effect; it is a phenomenon that only appears in relation to a system external to and sufficiently distant from the mass generating the gravitational distortion (what is commonly referred to as: observer). For the more adventurous, at this point we could embark on a journey through the meanders of the so-called *holographic principle*; perhaps it will be an opportunity to meet again in the future with some appendix and think together.

Before proceeding and taking a further step towards discovering the wonders of white holes, it is worth dwelling for just a moment on what it means that the event horizon is a perspective phenomenon – thus also making the discourse addressed in the first part of the text more complete. We cannot avoid using a metaphor, or if you prefer a thought experiment as it is customary to define these explanations in the world of theoretical physics. If we imagine that we, you and I, were riding a particle falling into a black hole, we would perceive no distortion. We could continue our discourse on the universe without perceiving ourselves slowed down; the clocks on our wrists would remain perfectly synchronised with each other. They would instead lag behind that of an astronomer friend trying to scrutinize us with his telescope from an earthly observatory.

A white hole – that was the starting point – is the symmetrical phenomenon to a black hole, a centre of mass into which matter-energy, instead of being unable to leave, cannot enter. More a source than a hole. Matter that *initially* falls into a black hole, attracted and trapped by its gravity, is compressed until it reaches the limit of the singularity. A concentration of mass whose density reaches the Planck scale. Under these conditions, the quanta assume a corpuscular behaviour and, becoming no longer compressible, produce a pressure from which a rebound originates. This point (which is also a momentum) is also called a Planck star.

The mathematics of this transition is well beyond the bounds of popularisation, but for our purposes it is significant to note that both loop quantum gravity and string theory can describe it consistently (i.e. without falling into contradiction). This is not proof that something like white holes "exist", but at least that they are not impossible according to our way of doing natural science.

Here is another crucial point, implicit in what has been said so far, but which it will be useful to make explicit. A white hole and a black hole, which now show themselves as two "moments" of the same dynamics of which they are both parts, share the same event horizon. Near these cosmic entities, our conceptions of time and space blur until they become irreducible to those we know in our world. From here, from our terrestrial observers, from *outside* the horizon, a white hole and a black hole are indistinguishable: gargantuan masses that attract what is nearby. Let us repeat: in the white hole one does not enter, in the black hole one does not leave; but an outside observer sees only the horizon of these celestial phenomena, and sees neither exits nor entrances, but only the approach of an object to the horizon on which it stops in a stasis, which from the outside looks surprisingly like eternity. On the inside, on the other hand, the transition – the *bounce* – could last a Planck time.

Only from the outside does time dilate, whereas if we assume the point of view of an observer inside the dynamics (an astrophysicist riding the bouncing black hole after reaching the floor of the Planck scale) would not experience any dilation. Dilation is in fact, as mentioned above, a perspective effect.

The story of this phenomenon can only be heard on the notes of mathematics, Galileo said almost half a millennium ago in The *Assayer*:

> Philosophy [this is how Galileo called what would later and thanks to him be called Modern Science] is written in this great book that is constantly open before our eyes (I say the universe), but it cannot be understood without first learning to understand the language and know the characters in which it is written. It is written in mathematical language, and the characters are triangles, circles and other geometric figures, without which it is impossible to humanly understand a word; without these it is a vain wandering through a dark labyrinth.

And the mathematics describing what happens inside the event horizon is that of the Einstein equations. From the horizon to the Planck star, they describe a black hole; from the rebound to the horizon, a white hole. But – we remind ourselves again – outside the horizon, in our world, nothing changes and what we see is always a huge, dense mass at the limits of possibility.

and its perspective effects. At only one point (or instant) are the equations of relativity violated: at the moment of rebound, when the scale of the phenomenon can only be explained by recourse to quantum mechanics. This is the point that is still being verified by researchers, but as anticipated, the results are positive under both the string model and loop quantum gravity.

Of what it is possible to measure and, at least in principle, experiment with, we have attempted to say. From here on, if you want to stay beyond the credits, we will try to peek at what tomorrow's science might be looking for. We will no longer be like wise dwarfs standing firmly on the shoulders of the giants who have gone before us. Rather, impatient and – I will not deny it – reckless, we will not resist the temptation to look further. We tiptoe forward to gain a more distant horizon and, catching sight of something intriguing, we will not shy away from a few leaps.

The elephant in the room of general relativity is that it smells of time travel. White holes add a piece to this theme that allows us to make the twin paradox less "mental" and use the mathematics behind it to describe the cosmos in which we live. This is where science comes from, from the desire to explain the unexplained; not to accept it as inexplicable but to investigate it, without rest or fear of the – inevitable – errors.

One fascinating point is that the mathematics describing a white hole is similar to the point of being superimposable on that of the Big Bang.

The history of relativity is a history of mental experiments, which later also turn out to be effective descriptions of the world. It is again Galileo who takes the first step in the *Dialogo sopra i due massimi sistemi del Mondo (Dialogue on the Two Greatest World Systems)* by proposing the image of a boat, which for a passenger below deck is indistinguishable whether it is stationary or "in uniform motion and not floating this way and that". Out of metaphor: two reference systems that are inertial to each other (i.e. at rest or in uniform rectilinear motion) are experimentally indistinguishable. This follows from the fact that time is assumed to be constant; and indeed, in our experience, time flows everywhere at the same rate (two identical hourglasses empty in the same time interval in our world). In Galileo's physics, which we call classical, the distance between two points understood as the length of a segment is considered invariant.

Narrow (or special) relativity formulated by Einstein in 1905 instead takes the speed of light as constant and invariant is no longer the distance between two points, but an interval that is both spatial and temporal. Time thus becomes the fourth dimension next to the three spatial dimensions of a single entity called *chronotope* or spacetime. This theory is also well described by a thought experiment: the twin paradox we have already mentioned. In short: at certain speeds close to that of light, time contracts. The theory is, like the Galilean theory, restricted to inertial reference systems.

General relativity was born by extending the same idea to those that can slow down or speed up. A thought experiment formulated by Einstein himself to explain the effects of the theory is that of the lift: if we found ourselves locked in an isolated room in free fall, dropping a handful of coins, they would float in mid-air, making the fall – ours and the coins – towards a centre of gravity such as the earth's, indistinguishable from the total absence of gravitational attraction. This is exactly the mechanism by which astronauts train for zero gravity inside planes that, having reached a certain altitude, are purposely brought to a stall and left in free fall for a few tens of seconds.

Theories with grandiose names; many efforts to construct metaphors, bridges between impeccable but often abstract mathematics and an experience as irreducible as it is evident – at least while we perceive it as sensation. What does not bear the brunt of this evidence is the fact – declared precisely by the theory of relativity – that space and time are not two distinct entities, two separate dimensions, but faces of the same coin (which seems to have four in all: three that we say are properly spatial and precisely one that is temporal). Another counterevidence is that this spacetime is as elastic and not rigid; not a constant as it seems to us from our sensations.

It is easier for us to imagine and thus accept this for space, because we experience rigid spatial objects and elastic ones. Elasticity is a reversible change, something that shows itself *over* time. Elasticity is a diachronic property and it is easy for us to imagine a space that – precisely over time – expands and contracts like a rubber sheet. It is – or at least it seems to me writing this – much more difficult to imagine this same property as an attribute of time. Almost as if it were a reflexive property. The form in which we translate (and inevitably *reduce*) this uncertainty into a language as rigorous as mathematics makes time as a denominator of itself. So, if the space that changes over time is velocity ($v = s/t$) while the change of the "whole" chronotype we approximate it to an acceleration ($a = s/t^2$). The exponent satisfies our intuition of the reflexivity of time. It seems less obvious to us than the way we write velocity as a ratio of "simple things", less intuitive; but it performs well when we use it in calculations. It returns useful results and that is why we continue to use it (our scientific paradigm is *pragmatist*).

What, then, is it legitimate, in the light of the Standard Model, to think of a white hole? *What lies* behind the numbers and formulas with which we describe it? Given that experiencing it directly seems very unlikely and that from our point of view we might not be able to distinguish "its body" from that of a black hole, what do we do with white holes?

One proposal, consistent with the mathematics that describes them, is to see them as *places in time*. Not unlike black holes in this, but necessary to complete the picture. Black holes are the places of time slowing down to the limit, white holes are the places (which are also moments since there is no space, but *space-time*) of time speeding up. Event horizons are the places/moments of light so far away or so slowed down by the mass that they appear still, immobile; in this sense eternal, as beyond that mode of being of time that we seem to grasp with the metaphor of the flowing river. And Planck's stars, the singularities, are places/moments of everything that appears to us as "first", and that is why we see them backwards in time and distant in space; because they are modes of being very different from the one we are in. Our laws, what in "our world" is invariant, such as the speed of light or the value of certain fundamental constants behave differently (in those places/moments, contexts, from our point of view would be quantified with different values).

On the wings of the hippogriff as the astronomer Schiapparelli liked to say, without contradicting ourselves, but without even hoping for some kind of feedback, we can think of the universe as a – at least – four-dimensional entity. Likewise, event horizons are four-dimensional entities in their turn; think of them as boundaries between different rhythms of the passage of time, "bubbles" inside and outside of which information cannot be transmitted bi-directionally. Wanting to venture a slightly more formal definition, a horizon is the set of phenomena that share the same gradient of dissipation of matter-energy and information. Then that dilation/expansion of space that we measure by scanning cosmic space, is perhaps conceivable as the effect of the interaction of what we call "the present" with what we call "the past" or "the future". An effect of the interaction between horizons. Particular physical conditions, such as the concentration of mass found in holes (whether white or black) slows them down and makes them potentially interact with distant places/moments when observed from our point of view and when measured with our constants (such as the speed of light).

How does it do it? What is the mechanism by which mass performs what to us appears to be a spell? Through the work of what we call gravity, a pressure is generated, a force so extreme that it shifts phenomena on the quantum scale; they stop being determinate things and become clouds of probability, which recondense – collapse – even in a place/momentum that from our vantage point looks like a jump in space-time.

However, this language does not seem to us to grasp the emotional and quasi-value scope of the question about the possibility of the universe being born from nothing. We would like something more. And here lurks a temptation. The wanting pushed beyond the boundaries of what is graspable. That nothingness does not allow itself to be grasped, barely allows itself to be said. If we try to throw out the bridge, and make the metaphor a simile, in the term of comparison lies the woundedness of the will that is pushed beyond.

Then we could say that the universe is an entity that transcends time. It does not become because it is itself time. Some images with an animist flavour: the universe as a river that feeds itself and whose waters are beyond flowing because they *are* their own flowing. Images.

Nothingness is the other from everything. The universe is born from nothing, in the sense that it is born *in* nothingness. In the nothing other than itself. The horizon of truth blurs.

This is still not science, but it attempts to sketch the contours of the possible. More than a mere suggestion; more than a myth but less than a formal hypothesis. It is an attempt to build a bridge from a point we see (our scientific paradigm: the Standard Model) to something that *might* nevertheless be part of nature. In the absence of any evidence to the contrary, so far, we are *free to believe* and by this driven to seek new wonders.

THE UNIVERSE BORN FROM NOTHING

Tiziano Cantalupi

CHAPTER I

AN EXCURSUS ON QUANTUM THEORY

"I will describe Nature's behaviour but if you do not like this behaviour, your learning process will be hampered. Physicists have learned to live with this problem: that is, they have realised that the essential point is not whether they like or dislike a theory [quantum theory], but whether it provides predictions in agreement with experiments [...] From the point of view of common sense, Quantum Electrodynamics describes an absurd Nature.
However, it is in perfect agreement with the experimental data.
I hope therefore that you will be able to accept Nature for what it is..."

Richard Feynman (Nobel Prize in Physics)

1. Preamble

Quantum theory, or quantum mechanics (or quantum physics), is a scientific discipline born to explain the fine structure of matter. Although it is, together with the theory of Relativity, the scientific paradigm of reference of the 20th century, it has never been able to surpass in significantly the narrow circle of insiders.

This fact is even more surprising in view of the fact that the most remarkable technological innovations, the most important scientific theories dealing with the indefinitely small or the infinitely large, are based on exquisitely quantum effects.

These effects relate to atomic energy, modern microelectronics (exploited in "classical" and quantum computers), digital clocks, lasers, superconducting systems, photoelectric cells, equipment for medical diagnostics and treatment, and many other applications in a wide variety of scientific and technological fields.

The reason behind quantum physics' isolation from the scientific and cultural landscape lies in the extreme conceptual complexity of its fundamental assumptions and the difficulty of its mathematical formalism, which make it a difficult subject even for physicists themselves.

Indeed, it is well known that the university exam in Quantum Mechanics is one of the toughest obstacles to overcome for any student in the Physics degree course.

Emphasis was thus placed on the conceptual complexity of quantum theory.

Actually, rather than conceptual complexity, one should speak of difficulty in accepting certain *counter-intuitive* postulates.

This feeling of inadequacy in accepting certain quantum assumptions was paradoxically also felt by the founding fathers of the quantum paradigm themselves (the so-called *Copenhagen School* exponents): Max Born, Niels Bohr, Werner Heisenberg, Wolfgang Pauli and, albeit separately, Paul A. M. Dirac.[1]

1　**Max Born.** Polish-born physicist (Breslau 1882 – Göttingen 1970). For his probabilistic interpretation of the "wave function", he received the Nobel Prize in 1954. **Niels Bohr.** Danish physicist (Copenhagen 1885 – 1962). Won the Nobel Prize in 1922 for formulating his model of the atom and the radiation it emits. **Werner Heisenberg.** German physicist (Würzburg 1901 – Munich 1976). He was awarded the Nobel Prize in 1932 for his formulation of the Uncertainty Principle. **Wolfgang Pauli.** Austrian physicist (Vienna 1900- Zurich 1958). For the discovery of the Exclusion Principle that bears his name, he was

The originator of the Uncertainty Principle, Heisenberg, expressed himself on this subject as follows:

> I remember discussions with Bohr that went on for many hours into the night and led us almost to a state of despair, and when I would leave the discussion and go for a walk in the nearby park, I always kept asking myself the question: is it possible that nature is as absurd as it appears to us in these atomic experiments [in these quantum physics experiments]? (Heisenberg, 1961).

1.1. Fundamentals of quantum mechanics

- There is no definite reality of matter, but an objectively *indistinct reality made up of superimposed states.*
- The fundamental dynamics of the microworld are characterised by acausality and the "nonclear" separation between experimenter, measuring apparatus and observed object. Whether the matery manifests itself in a definite way in one form rather than another depends only on the type of experiment chosen to observe it.
- It is possible that, under certain conditions, what happens in a given place may have an instantaneous counterpart in a place far away from it.
- Matter and energy can (for a very short time) spring from the vacuum.

The list just given of the fundamental assumptions of quantum mechanics makes us realise how difficult it has been (and still is) not only to accept, but also to explain, the foundations of this theory

---

awarded the Nobel Prize in 1945. **Paul A. M. Dirac.** British physicist (Bristol 1902 – Tallahassee 1984). For his research into the theory of the anti-electrone (the positron) he was awarded the Nobel Prize in 1933.

without running the risk of not being understood, or, even worse, of being misinterpreted. The language available to physicists or anyone else in the field who tries to express concepts such as acausality or superposition of states is very often inadequate. The words that languages make available to us to communicate certain concepts, certain experiences (or to organise a coherent gnoseological scenario), are often inadequate, not least because words are designed to describe and present *ordinary reality*, but quantum mechanics has very little that is *ordinary*. This aspect of the "describability" of quantum phenomena is well expressed in a sentence written by Max Born (1957):

> The ultimate origin of the difficulties lies in the fact (or philosophical principle) that we are forced to use words from everyday language when we want to describe a [quantum] phenomenon. Common language has grown out of everyday experience and can never go beyond certain limits...

1.2. Classical physics

This section is dedicated to the classical view of science and is dedicated to the physics that preceded the advent of quantum mechanics. In ancient times, people who began to question the dynamics of natural events happening around them received a very "fleeting" image from the world around them. They realised that some events were regular and predictable; the alternation of day and night, the cycle of the seasons, the phases of celestial bodies, while others were occasional and appeared irregular, such as weather events, earthquakes and volcanic eruptions. How could those men have organised their knowledge into an explanatory framework of nature? In some cases, for natural events they found an obvious explanation: observing, for example, snow melting in the heat of the sun or fire extinguishing on contact with water.

However, the concepts of cause and effect were not clearly understood and represented: instead, it must have been natural to construct a model of world events based on the intervention of supernatural forces. The most advanced societies constructed a highly complex and highly anthropomorphic hierarchy of deities. Earth, Moon, Sun, planets were regarded as human-like characters and their events as the equivalent of human actions, emotions and desires. "The wrath of the gods" could thus be seen as a sufficient explanation for natural calamities, to be appeased by appropriate sacrifices.

Parallel to the establishment of the above-mentioned ideas, another set of beliefs grew up following the development of cities and the appearance of states. In these contexts, it was thought that, in order to avoid confusion, citizens should conform to a strict code of behaviour, which was institutionalised in the form of laws. As was to be expected, the gods too were considered subject to laws and, by virtue of their greater authority and power, sanctioned the system of human laws through their intermediaries, the priests.

In ancient Greece, the conception of a universe governed by laws was already widespread. Explanations for events such as the fall of a stone or the flight of an arrow were already formulated in the form of unalterable *laws of nature*. This clarifying conception of phenomena, which would occur independently in close relation to natural laws, was in stark contrast to the conception of an overall finalistic world. Of course, the really important phenomena, such as the creation of man and the universe or the cycles of the stars, still required the strictest "attention" of the gods; *everyday events* could go it alone. But once the idea took root that a physical system could evolve autonomously according to certain fixed and inviolable principles, a gradual erosion of the gods' dominion was inevitable.

Although the renouncement of the theological interpretation of the physical world is not complete even today, one can roughly link the decisive turning point in favour of the physical world to Galileo Galilei,

Isaac Newton and partly to Charles Darwin, the decisive turning point in favour of the importance of physical laws. At the end of the 16th and 17th centuries, what is now unanimously referred to as the father of "modern science", Galileo, began a series of experiments destined to change the course of physics and beyond. The central idea of the genius from Pisa was that a piece of the world as isolated as possible from external influences would be relatively easy to study, while at the same time demonstrating a relatively simple com- pact. One of the best known investigations conducted by Galileo was the observation of the behaviour of falling bodies. The pre-falling of a body is usually a very complex process, dependent on weight, shape, composition, wind speed and air density. Galileo's genius lay in realising that all these variables are merely accidental complications superimposed on a law that is actually very simple. By rolling rectangular-shaped bodies along inclined planes, which drastically reduced the influence of air, Galileo was able to overcome the complexity (accidental perturbations) and isolate the fundamental law of falling bodies. Today, the procedure adopted by Galileo may seem most reasonable, but in the 1500s it required a considerable amount of inventiveness, just as a considerable amount of ingenuity required the introduction of the concept of time in the study of motion.

By introducing the time variable in the study of falling bodies, Galileo discovered a law as simple as it is fundamental, namely that *the time required to fall from a given height (starting from a position of stillness) is exactly proportional to the square root of the height stressed*. With this simple relationship, "modern science" was born. The idea of a mathematical formula instead of a regulating god to describe the behaviour of a physical system made its appearance. The importance of this intellectual breakthrough for the progress of mankind cannot be overemphasised. A law of nature expressed as a mathematical equation means not only synthesis and universality, but also the possibility of calculation. This meant that it was no longer essential to observe nature in order to ascertain its behaviour (or evolution): it could also be calculated at a desk with pen and paper. By using mathematics to explain the laws, scientists could predict the behaviour of nature (or evolution). Future shape of the world, as well as reconstruct the past one.

At the end of the 17th century, the English physicist Isaac Newton took up and developed Galileo's work by elaborating principles generally applicable to all moving bodies. Generalising the results obtained by Galileo for the force of gravity on Earth, Newton hypothesised that the Sun and all bodies in the universe exerted a mutual gravitational force that decreased with the square of the distance (once again, we are dealing with a simple yet precise mathematical relationship that calls into question the inverse of the square).

Having *mathematised* motion, Newton also *mathematised* gravity. By combining the two, and using infinitesimal calculus, he achieved a great result by accurately predicting the behaviour of the planets and the dates of eclipses.

The extension of Newtonian mechanics to the solar system was more than just an exercise in applied physics and mathematics: it de-molested the centuries-old belief in the idea that the dynamics of the cosmos were governed by purely *celestial* forces.

Newton's universe is a perfect mechanism. The movement of every grain of matter, of every atom, is in principle completely and absolutely determined, for all future and past time, by knowledge of the forces imparted and the initial conditions. Knowledge of the initial conditions and the possibility of "calculating" future events (or effects) strongly characterises Newton's work, as does all of science until the advent of quantum mechanics.

1.3. The Discovery of the Quantum

The year 1900 officially saw the birth of quantum mechanics. This was the year in which the German physicist Max Planck discovered that all energy manifestations, from the radiant flux from a heat source such as a burning piece of wood, to the light produced by the sun, are transmitted by means of exchanges of discrete entities. Planck, in particular, realises that in order to adequately explain what was known in physics as "the problem of blackbody radiation", it was necessary to abandon the assumption of classical physics that the emission of energy is continuous and replace it with the hypothesis (shocking for the time) that energy must be delivered in discrete, discontinuous quantities. Planck called these discrete quantities *quanta*, plural of the Latin word *quantum*, which literally means "how much" (hence the term quantum mechanics) and represented them as the *action quantum* identified with the letter h (h is a very small number corresponding to a magnitude of $6,6 \cdot 10^{-27}$).

The success of Planck's hypothesis had, however, problematic aspects: to admit that radiant energy could only be discretely emitted or absorbed was to implicitly recognise that the energy in a light wave was not continuously distributed but was concentrated in the form of light grains or light corpuscles. Planck, who was a conservative and, at forty- two years of age, was, according to the criteria of science at the time, already considered old, at first rejected the disruptive scope of his ideas, fearing that the wave structure of light (perfectly described – at least in appearance – by Maxwell's electromagnetism) might be called into question, but all experiments confirmed[2] his hypotheses.

---

2   The most famous experiment confirming the reliability of Planck's hypothesis involves Einstein's photoelectric effect. It is curious to note that, contrary to what many people believe, Albert Einstein did not win the Nobel Prize for the formulation of the theory of Relativity, but for the discovery, or rather, for the explanation of the photoelectric effect first highlighted in 1887 by H. R. Hertz. By explaining the photoelectric effect, Einstein (who also introduced the concept of the ***quantum of light*** – today's photon) was able to account for the fact that a sheet of metal, when hit by a beam of light, emits a cloud of electrons. According to the German physicist, the reason for this emission is to be found in the discrete nature of light; it is to be found in the interaction between the incident particles of light (the photons) and the electrons of the metal atoms, which are knocked out of their orbits when they are hit.

At a certain moment, Planck, also pressurised by the facts, ended up accepting his own ideas. From then on, the discontinuity of energy phenomena saw an extension to every field of physics and beyond.[3]

The *how much action h* ended up becoming a fundamental constant of nature (the well-known "Planck constant"), in the same way as Einstein's c (the speed of light in vacuum).

1.4. The wave-corpuscle dualism

With the discovery of quanta, physics in the early 20th century was faced with a profound contradiction. On the one hand there were the works of Planck (confirmed by Einstein and Compton), which demonstrated the corpuscular nature of light phenomena, and on the other hand there were the works of Young, Fresnel and Maxwell (formulated in the 19th century), which indicated that light could only be a wave phenomenon. In particular, Young's *interference* experiments unequivocally demonstrated the wave-like nature of light phenomena.

The possibility of analysing and understanding the phenomenon of *interference* is given to us by the observation of the water of a pond where two bodies are thrown simultaneously. Each of the two bodies initiates a sequence of ripples moving away from the impact area. At the point where the two series of ripples meet, a systematic pattern of ridges and troughs forms on the water surface. This is due to the fact that where the wave ridges of one series coincide with those of the other, the disturbance is reinforced, while where the ridges of one series meet the troughs of the other, the two disturbances cancel each other out and the water surface is almost flat.

---

3   The corpuscular nature of "energy phenomena" also had repercussions at a philosophical level. The metaphysical principle of the **continuity of natural phenomena** expressed in the famous phrase: **natura non facit saltus** and strongly supported by, among others, the great Leibniz, was called into question following the discovery of quanta.

If one now imagines replacing the wave motions produced by the impact of the pair of stones on the water by two *streams of* light waves (coming from two parallel directions) which end their course on a screen, one will see that as a result of the interaction between the waves on the screen, light and dark bands will appear from the interference of the single wave amplitudes. Now, in the case where two crests are "in phase", there will be a reinforcement and the resulting amplitude will become double (*constructive interference*). In the case where a ridge and a re-entrant are opposite to each other, they will cancel each other out, with consequently zero amplitude (*destructive interference*). The figure formed on the screen as a result of the interaction between the light waves will therefore consist of a series of light bands, in the case of *constructive interference*, and dark bands, in the case of *destructive interference*. This experimental situation demonstrates (as well as many other contexts in optical physics) unequivocally the wave-like, continuous nature of light.

Here, then, the beginning of the 20th century saw physicists attempting to interpret the nature of light phenomena faced with a strange dilemma. Depending on the mode of observation, light appears as a corpuscle (photoelectric effect) or as a wave (*interference* effects). It should be clear to anyone the absurdity of such a situation: especially if one considers that a wave is something that occupies volume, whereas a corpuscle is something concentrated at a precise point in space. Depending on the experimental situation, therefore, light manifests intrinsically opposite characteristics.

1.5 Matter's undulatory nature

In 1923, the French physicist Louis de Broglie (meditating on the symmetries of nature) put forward a daring and ingenious hypothesis: why, he asked himself, since light manifests itself in two aspects, wave and corpuscle, could it not be the same for matter as well? He therefore formulated a theory, which associated to each corpuscle of matter a wave of given length λ ($\lambda = h/m \cdot v$: where h is Planck's constant, m and v are, respectively, the mass and velocity of the moving corpuscle), i.e. a periodic phenomenon extended in the space surrounding the particle. The dualistic nature of light phenomena was thus extended by de Broglie to every form of matter: from the small electron, to the atom, to any macroscopic entity in motion.

By carefully observing the above formalism (and not forgetting the extreme smallness of the value of the quantum of action h), it will be understood how, with the same velocity, the mass of the moving entity considered is decisive for the definition of the wavelength associated with it. The larger the mass, the shorter the length of the wave. This explains why matter waves cannot be relevant in the macroscopic world. In fact, de Broglie's equation tells us that such waves, even when associated with one of the "smallest" visible objects – a sand granule – are so small compared to the size (mass) of the object that their effect is negligible. Only at the level of subatomic particles does the wave nature of matter become relevant. The very small size of an electron, for example, implies that this particle is greatly influenced by its associated wave.[4]

Following de Broglie's work, therefore, a beam of light, a beam of electrons, can be imagined either as an electromagnetic "train of waves" or as a "jet of balls" moving through space.

At the beginning of 1926, the Austrian physicist-mathematician Erwin Schrödinger formulated an equation that allows a complete

---

4   In the following, we compare the associated "size"-wavelength relationship of a macroscopic entity and a microscopic entity. The entity from the macroscopic world considered is a fly. An average-sized fly has a mass of about 1 g and can fly at up to 100 cm/sec. Its associated wavelength will therefore be: $\lambda = h/m \cdot v = 6.6 \cdot 10^{-27} /1 \cdot 100 = 6.6 \cdot 10^{-29}$ cm; the 10 raised to the -29 cm makes us realise how small the wave associated with a fly. Let us now look at the magnitude of the wave associated with an electron. An electron has a mass of $9 \cdot 10^{-28}$ g and if it moves at a speed of 100,000,000 cm/sec it will have an associated wave (according to the formula $\lambda = h/m \cdot v$) of about $7.3 \cdot 10^{-8}$ cm.

description of every single wave property of matter. It improves on de Broglie's material-wave equation, and with the introduction of the *wave function* or *wave equation* (identified with the Greek letter Ψ), it makes it possible to describe every single behaviour in space and time of the entities in the microworld, as well as to calculate the distribution of the "statistical possibility" of finding a particle within space corresponding to the dimensions of its associated wave.

1.6. Confirmation of matter's undulatory nature

Around the mid-1920s, the American physicist Clinton Davisson was conducting research for the Bell Telephone Company when, in the course of certain experiments, he observed some curious deflections of electrons upon interaction with nickel crystals, the nature of which he could not explain. In 1927, Davisson performed an improved version of the experiment together with a colleague, Lester Germer, and realised that these strange deviations were nothing more than "diffraction figures" typical of the reflection of waves – in this case "electron waves" – on nickel crystals. Subsequent experiments not only confirmed Davisson and Germer's results, but also showed that atoms and molecules also exhibit wave-like behaviour. Thus, practically any "object" has an associated wave: an electron, a golf ball, a car and even a person, although for the latter it is insignificant in size and influence.

1.7. Probabilism and acausality

At the beginning of the 20th century, physicists believed that all processes in the universe were perfectly calculable as long as sufficiently precise starting data were available. This deterministic philosophy, as we have seen in the section on *classical physics*, had started more than two centuries earlier when Newton had succeeded in describing the orbits of the planets with his law of universal gravitation. In one fell swoop, the English scientist had shown that an apple falling from a tree and a celestial body moving through space are governed by the same law: the universe ticked like a giant, perfectly regulated clock. The 19th century French mathematician, Laplace, was one of the staunchest advocates of determinism. On more than one occasion, he did not fail to emphasise *that if an omniscient intelligence* (a sort of *superman*) were able to observe all the forces at work in nature and record the position of each fragment of matter at a particular time, he would be "able to include the motions of the largest bodies and those of the smallest atoms in a single formula [...] nothing would have resulted indefinite; in his eyes, future and past would become present".

But coinciding with the end of the Victorian era, the belief in a perfectly "calculable" and comprehensible universe vanished; it happened as physicists attempted to apply deterministic laws to the behaviour of the atomic world. In that tiny realm, as we have seen, matter seems to revel in manifesting contradictory aspects.

At the end of 1926, the physicist Max Born, partly in the light of the impasse in which physics had found itself, came up with a profoundly innovative thesis that gave a quantum-mechanical interpretation of the relationship between the wave and corpuscular aspects of matter. Born's hypothesis, which in fact transformed the Schrödinger's "statistical possibility" of finding a particle within its associated wave, into a "probability", was a disruptive idea and generated a very long series of discussions, as it sanctioned the end of determinism in physics. This hypothesis was later accepted by most physicists as being in agreement with experimental results.

In classical theories, the calculation of probability is essentially governed by practical reasons, not by principle. According to Born's quantum interpretation, on the other hand, the exact result of an experiment is generally unpredictable even in principle, and probability plays the role of a primitive notion imposed not by ignorance of causes that can be known exactly, but by the fact that causal analysis is limited by the fundamental laws governing the phenomena of the atomic world. Bornian probabilism stems essentially from the wave-corpuscular behaviour of matter and the fact that this behaviour does not allow unambiguous and certain predictions to be made about the evolution of microsystems. Schrödinger's wave equation tells us that a particle can occupy all possible positions within the associated wave. By occupying all possible positions, the particle no longer has a defined location and trajectory: the lack of these "conditions" no longer makes it possible to make precise predictions about its future behaviour. Classical mechanics admits that it is possible to make deterministic predictions only in the case in which contemporaneous information is available on the canonical co-ordinate values (position, "velocity") of the object under examination at a given instant. The probabilism highlighted by Born, (together with the indeterminism inherent in the Principles formulated by Heisenberg; we will see later what this is about), emphasises the impossibility of knowing with precision the canonical starting co-ordinates of any micro-object, and therefore precludes the possibility of making certain predictions about the future behaviour of any micro-object, as well as of all the "objects" that interact with it.[5] The impossibility of making certain predictions about the spatio-temporal evolution of the entities of the atomic world undermines the principle of causality.

To summarise what has been said so far, it can be stated that around the mid-1920s, Max Born proposed a probabilistic interpretation of

---

5   In an article in issue 302 (1993) of the journal "Scientific American", the American physicists Chiao, Kwiat and Steinberg give a clear example of how a particle, in the quantum view, loses a definite location and trajectory forever: "In quantum mechanics, the concept of a trajectory breaks down: the position of a particle is not described as a precise mathematical point; instead, the particle is better represented as a wave packet [the term wave packet is synonymous with the wave function ψ] spread. Therefore, when the particle is revealed at a point, the entire wave packet disappears. Quantum mechanics does not say where the particle was before that moment".

Schrödinger's wave function Ψ. To explain what the wave must be and what the particle must be, he constructs a representation in which Ψ represents the *probability distribution of* finding a particle at each point in space formed by the associated wave. Born compares the material wave originally postulated by Schrödinger to the shock wave produced by an explosion: its density must be very high near the point of deflagration and must decrease as one moves away. The figure below visually represents the shape of the probability wave postulated by Born.

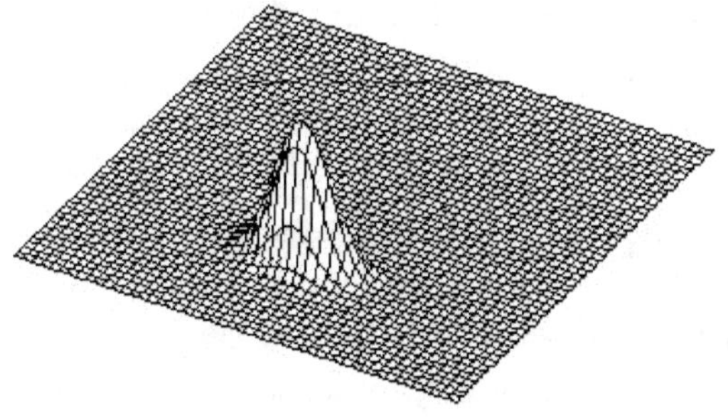

In the Bornian paradigm, the electron (or any other microentity) forever loses its specific location. Of an electron revolving around a nucleus of a hydrogen atom, for example, it can only be said to have a certain *probability* of being found at a given point in space (of the orbital level) around the nucleus itself.

The Italian physicist Franco Selleri, in his book *La Causalità Impossibile,* offers a clear example of the meaning of quantum probabilism and the acausality that follows; he writes:

> The problem that arises very naturally is to understand the causes that determine the different individual lives of [free] neutrons. The same problem arises for any kind of unstable system such as excited atoms [...]. The Copenhagen interpretation of quantum theory not only provides no knowledge of these causes, but explicitly accepts an acausal philosophy according to which every disintegration process of an unstable system has an absolutely spontaneous [probabilistic] nature that admits

of no explanation in causal terms. According to this line of thought, the problem of the different individual lives of unstable systems should necessarily remain unanswered and should indeed be considered an unscientific problem. (Selleri, 1987)

It was following the introduction of quantum probabilism and its consequences for cause-effect relations that Einstein uttered the famous sentence: "It seems difficult to take a look at the cards God holds in his hands, but not for a moment can I believe that he plays dice".

And again, in a letter sent to Born, Einstein writes:

> The [quantum] theories of radiation are of great interest to me, but I would not like to be forced to abandon strict causality without defending it more tenaciously than I have done so far. I find absolutely intolerable the idea that an electron exposed to radiation chooses of its own free will not only the moment to "jump", but also the direction of the "jump". In that case I would rather be a casino dealer than a physicist.

From the quotations just given (and those to come on the following pages), Albert Einstein's critical stance towards quantum mechanics is evident. This position appears all the more peculiar when one considers the great German scientist's considerable contributions to the development of the quantum paradigm. One thinks of the explanation in terms of the interaction between light quanta and atomic electrons of the photoelectric effect. One of Einstein's most influential biographers about the troubled relationship between the father of relativity and quantum theory writes: "No physicist has contributed more to the creation of quantum physics than Einstein. His work in this field would in itself have meant an entire scientific career for any other physicist [...] he no less decisively rejected it when it was now generally accepted".

From a strictly *human* point of view, the opposition between Einstein and the exponents of orthodox quantum theory (the *Copen-*

hagen School exponents) sees the continuous struggle of a man who proceeds, surely and self-consistently, along a road far from the main course of physics, but in his opinion the only true one for understanding the mechanisms of nature.

1.8. The superposition of states and the act of observation

Born's probabilistic interpretation of the Schrödinger equation constitutes a special aspect of quantum theory, since with it the rigidly deterministic laws of classical mechanics are replaced by probabilistic laws. Within quantum mechanics, the probabilistic interpretation of ψ gives rise to a further peculiar aspect: the "superposition of states". A psychological analogy will be used to explain what this is.

Suppose that a person's mood is either Bad (state C) or Good (state B) and that the probability of finding them in either of these two states is 50 %. The quantum mathematical formalism then tells us that the mood of the person in question at any given time of the day is represented by the linear superposition of sub-states C and B, but that the probability P of finding her in a bad or good mood is in the ratio 0.5 to 0.5, which translates symbolically as: $P = 0.5\,P\,(C) + 0.5\,P\,(B)$. We can therefore say that the person's mood fluctuates from state C to state B, and in order to find out more, we need to meet her ("observe") to see what her mood really is. We will be able to find her in state B, an experience that will allow us to say that at the time of the encounter/measurement we have *reduced her wave function* to only *the* sub-state $P = (B)$: state P is also called *the state vector*.

Quantum mechanics offers very significant examples of over- position of states; one of these is the lifetime of the meson K^0 (the meson is a sub-atomic particle). The lifetime of K^0 can be 10^{-7} or 10^{-10} seconds. Applying the superposition of states' theory, it can be stated that the lifetime of K^0 is the superposition of two states K^0 (10^{-7}) and

K^0 (10^{-10}). The meson can thus, in equal parts and depending on the moment in which it is observed, *probabilistically* manifests itself after a certain time. This is exemplified by the formula $| K^0|^2 = 0,5 |K^0|^2 + 0,5|K^0|^2$. The fact that the probability is the square of the coefficient that determines its influence at the moment of the reduction of the wave function, is due to the mathematical formalism of quantum mechanics (owed to Born), which requires the effective probability to be the square of the probability amplitude.[6]

Another example of quantum superposition concerns the dual behaviour of matter. As pointed out in the preceding paragraphs, matter manifests a dual nature: undulatory and corpuscular. Well, these two characteristics of physical reality – in the quantum paradigm – must coexist together in a mixture of superpositions: in this case superpositions of wave and corpuscular states. It is the type of measurement chosen by the experimenter (which produces the *reduction of the wave function*) that determines how matter is to present itself to the world; whether as wave-like or corpuscular.

From what has been written so far, the reader will have realised that the "concretization" of the characteristics of matter, in particular in the case of the wave-corpuscle superposition, does not occur "spontaneously" as in the life-span of the K^0 meson (which "spontaneously chooses" when to manifest itself: whether after 10^{-7} or after 10^{-10} seconds), but is "forced" by the experimenter's choices: we will see what this entails later.

To visually exemplify a state of overlapping, one can turn to the *doubly ambiguous* images characteristic of the German Gestalt psychological movement. In the contours that constitute Gestalt figures,

---

6   When calculating the sum of the squares of the probability amplitudes of the various states in certain formal situations, these amplitudes must be summed up as the square of a binomial. This procedure leads to what in quantistic mathematical formalism are called "crossed terms", which lead to non-localism (we will see what this is in the next Sections).

the two meanings assumed by them are simultaneously present. The figure below is concretely and at the same time the representation of a vase and two human profiles.[7]

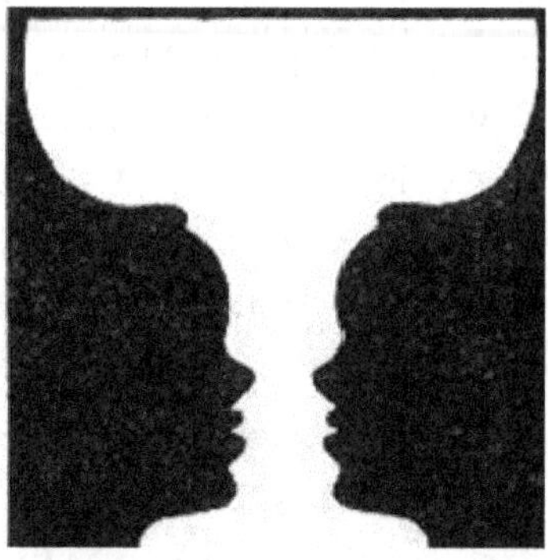

At a certain point in the exposition concerning *overlapping states,* the circumstance was explicitly mentioned that knowledge of a person's mood can only occur following an encounter (to an "observation"). This active aspect of the knowledge process is fundamental in quantum mechanics. Within the Copenhagen paradigm, in fact, the role of the experimenter, of the observer (in quantum physics, the person who performs any kind of experiment is defined as the *observer*), apart from being ineliminable, can never be separated from the observed entity. Observer and object being studied form a *whole at* the time of a measurement, so that the choices of the former determine and merge with the characteristics of the latter. All this derives

---

7   For the sake of truth it must be stated that for the exponents of orthodox quantum mechanics it is impossible to provide any representation of a superposed state. A superposed state is in itself unrepresentable, unthinkable: "It is impossible to understand the structure and evolution of a superposed state in the sense of forming mental images in space and time corresponding to its reality," writes Bohr in one of his best-known works. The reason why the writer has decided to depart from this "law" of quantum theory is to provide some kind of representation of a reality (superposed reality) that would otherwise elude any mode of representation.

substantially from the fact that matter, prior to any measurement (matter in its "natural state") *lives* in a superposed state. It is only the intervention of the experimenter who, by means of an act of observation, produces what in quantum formalism is called a *reduction of the wave function*, allowing the superposed state to resolve itself in a certain way, forcing matter to present itself to the world as an "entity" that has a certain "characteristic".

The interaction between experimenter and observed object, in the case of the life of the K^0 meson, "forces" the meson to tell us something about the duration of its life, while in the case of the wave-corpuscular aspect of matter, the observer-object-observed interaction forces (depending on the type of experiment chosen) matter to present itself to the world in a certain form: i.e. as a wave or as a corpuscle.

Within the quantum paradigm, therefore, the experimenter is a *participating observer*, he is an "element" that is never truly detached from the object of investigation. The measuring apparatus that intervenes in the experiment and acts as an intermediary between observer and observed object, also becomes an *active entity* in the measurements. In this regard, Bohr, in his book *Unity of Knowledge*, writes: "The fundamental difference, with respect to the analysis of phenomena, between classical physics and quantum physics is that in the former, the interaction between objects and measuring apparatus can be neglected or eliminated, whereas in the latter, this interaction is an integral part of the phenomenon".

1.9. The superposition of states and the double-slit experiment

There is a test in quantum physics that more than any other highlights the superposition state that characterises matter "in its natural state", this test is known as the *double-slit experiment.*

Let us assume, as in the example proposed to explain *interference* in the section on wave-corpuscle dualism, that two light-wave beams from two parallel directions are produced, which terminate on a light-sensitive screen (a kind of photographic plate). In the experiment illustrated in Figure 1 (which is a typical example of a two-slit experiment), parallelism between the light beams is achieved by opposing a light source to a diaphragm with two slits.

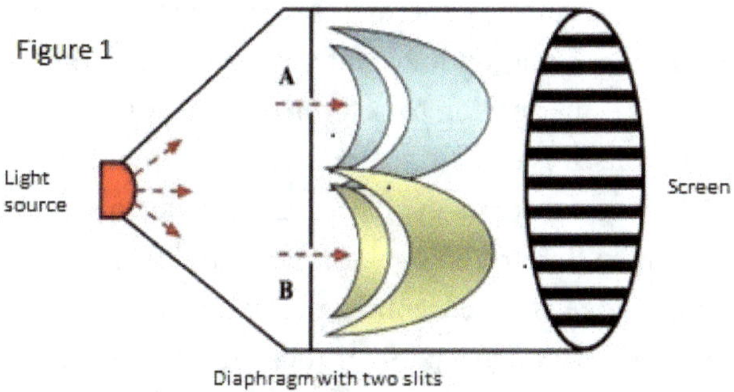

Now, as the figure clearly shows, in the central part to the right of the diaphragm, the waves of the light beams coming out of the slits interfere with each other, and if these waves are "in phase", a light band (*constructive interference*) appears on the screen; if these waves are "in opposition", a dark band (*destructive interference*) appears on the screen. According to the laws of optics, the interference just described can only take place if both slits are crossed by the light beams. In the case where only one of the slits is through which a beam of light passes – there being no interference – no band can be produced on the screen. The extraordinary thing that happens, however, is that even if only one beam of light is sent – or rather, even if only one photon is sent at a time – light and dark bands typical of constructive and destructive interference still form on the screen. How can this happen, physicists at the beginning of the 20th century asked themselves in puzzlement. The answer came from quantum mechanics and in particular from the work of Born, Bohr and Heisenberg, who demonstrated that any "micro-system" (photon, electron, atom) is not obliged by deterministic laws to follow precise trajectories. Bornian "probabilism", together with the superposition principle and Heisenberg's indeterminism, explicitly forbid any micro-being to possess a definite trajectory. In the case of the two-slit experiment illustrated in Figure 1, even a single photon therefore traverses all possible ("potential") "superposed" trajectories between the source and the aperture. Travelling through all the possible trajectories, the photon inevitably also encounters the two apertures, succeeding, after having passed through them (and thus having interfered with itself), to produce on the screen the light and dark bands typical of constructive and destructive interference. In a famous passage written by Bohr on this subject we read: "When the 'photon' passes through slit A this determines a certain possible world (which we shall call world A); when it passes through slit B, we shall have world B instead. In our case it means that both these worlds coexist in some way, one superimposed on the other". And again the physicist Richard Feynman writes: "In a two-slit experiment, the micro-object does not pass through only one of the two slits. The quantum-mechanical explanation of the double-slit experiment was greeted (when it appeared) with scepticism by part of the scientific community." Several physicists, especially those who more than others subscribed to the realist position,[8] did not believe admissible that an entity as special as a photon, an electron or an atom could "split" and travel different paths at the same time. To be certain of the correctness of the quantum positions, a simple yet decisive experiment was then proposed.

---

8   Realism is that complex of scientific-philosophical positions advocating the t h e s i s that "objects in the world" always have a defined reality (e.g. a precise position and speed), which does not depend on their being perceived (or measured). Realism is also based on the unconditional acceptance of "localism", i.e. the impossibility of "direct actions at a distance" and causality, i.e. the evidence that all effects always propagate from the past to the future and that the past cannot be changed.

If indeed a single photon passes through both slits, self- interacts with itself and produces the typical interference bands, it is sufficient to close one of the two slits and verify if indeed the bands continue to occur. The experiment was performed, only one photon at a time was sent into the measuring apparatus, and the outcome was crystal clear: closing one of the slits blocked the formation of the light and dark bands on the screen. Although it may be difficult to believe that a single object can have the ability to "probe" two different places at the same time, this is what happens, this is what the experiments show.

1.10. The Principle of Uncertainty

In 1927, the German physicist Werner Heisenberg discovered that the probabilistic nature of the laws of quantum mechanics placed great limits on our knowledge of an atomic system. Normally, one expects the state of a moving microparticle – consider, for example, an electron rotating around a nucleus of an atom – to be characterised completely using two parameters: velocity (more precisely, momentum) and position. Heisenberg postulated instead, that at some level these quantities (which in technical language are called *conjugate variables*) should always remain undefined. This limitation took the name of the Uncertainty Principle. This principle states that the greater the accuracy in determining the position of a particle, the lower the precision with which its velocity can be ascertained, and vice versa.[9]

When thinking about the equipment needed to carry out measurement rations, this uncertainty is intuitive. Detection devices are so large in relation to the size of a particle that the measurement of a parameter such as position is bound to change the velocity as well.

---

9   More precisely, in the simultaneous measurement of the position coordinates x and velocity (momentum) p of a particle, it is impossible to obtain x' and p' values with any small uncertainty. In fact, if Δx and Δp define the uncertainty in x and p respectively, the relation: $\Delta x \cdot \Delta p = h$ must hold.

However, it must be emphasised (following Heisenberg) that the limitations in question do not only arise from the interaction between the microscopic and macroscopic worlds but are intrinsic properties of matter. In no sense can it be assumed that a microparticle possesses a certain position and speed at a given instant. These are, according to Heisenberg, incompatible characteristics; which of the two manifests itself more precisely depends only on the type of measurement chosen. In addition to the position and velocity of particles, the Uncertainty Principle also places limits on the simultaneous measurement of parameters such as energy and time. This means that as the duration of an observation decreases, the inaccuracy in the measurement of energy increases, and vice versa.[10]

Finally, it should be noted that the Uncertainty Principle, as well as the relations that bind the wavelength dimension associated with a moving material entity theorised by de Broglie, is valid for any "object", but in practice only has important consequences when applied to objects of atomic or subatomic dimensions, since when it comes to ordinary bodies, the very small size of the constant h, causes the principle itself to lose much of its influence.

1.10.1. Experimental uncertainty

In order to measure the position of a microscopic object such as an electron, it is necessary to hit it with a beam of light (a stream of photons), or something that ultimately turns out to have approximately the same *size/energy* as the electron. This causes the electron to be perturbed by this interaction, which inexorably changes its speed;

---

10   More precisely, energy-time defines the impossibility of determining the energy of a particle with an uncertainty less than ΔE, linked to Δt by the relation: $\Delta E \cdot \Delta t = h$.

The same thing, but in opposite situations, occurs in the case of wanting to know the velocity of an electron. Heisenberg, with his Principle of Uncertainty, strongly emphasises the fact that the effect of the interaction between the incident light beam and the "observed" particle can also be made equal to zero, but in this case no measurement is performed and no knowledge of the properties of the observed system is gained. If, on the other hand, the interaction is non-zero, it cannot be made arbitrarily small and is therefore not conceptually eliminable either. An example of experimental uncertainty brought down to human dimensions may be the attempt to simultaneously determine the trajectory and position of a moving object. Suppose we want to determine the position of a moving object, for example a cannonball, by means of a photograph. Naturally, if the ball travels along its trajectory at considerable speed, the photograph will be blurred unless a high-speed shutter is used. With a high-speed shutter, you will be able to "fix" the position of the object on the film, but you will pay a price in defining the trajectory, resulting in a stationary cannonball. On the other hand, using a low-speed shutter will photograph an indistinct line that faithfully represents the trajectory of the ball but gives no indication of position.

1.10.2. Theoretical Uncertainty

As noted above, the fact that the position and velocity of a particle cannot be measured temporarily is not merely due to practical restrictions but is an objective limitation of nature. In other words, for Heisenberg, and for quantum physicists in general, any micro-entity (matter) in its "natural state" does not objectively (ontologically) have a defined position and velocity, it does not have a "defined reality". This is a consequence of the superposition principle and the mathematical formalism underlying uncertainty relations. In fact, as has been pointed out so far, matter, prior to any measurement, always *lives* as a mixture of superposed states. In the case of variable conjugate velocity-position and energy-time, the particle *lives* as an entity that has these four properties "mixed". The analysis of the formalism at the basis of the position-velocity uncertainty relations, $\Delta x \cdot \Delta p = h$, with the "inverse" equations $\Delta x = h/\Delta p$ and $\Delta p = h/\Delta x$, shows that if, for example, we wish to precisely define the position Δx of a particle, its velocity Δp must tend to zero, but if Δp tends to zero, Δx will tend to infinity and vice versa. The simple example above clearly demonstrates the incompatibility for a particle to possess (except as a "mixture" of indistinct, superimposed states) at one time position and velocity: when one of these two variables tends to zero the other tends to infinity and vice versa. With the theoretical demonstration of the uncertainty relations, the inevitable passage from the epistemic level of experiment, where instrumental apparatuses show insurmountable limits in the possibility of carrying out certain measurements, to the ontological level of formalism, where matter demonstrates that it does not possess certain characteristics in principle, becomes clear.

1.11. Complementarity

The *contradictory* aspect inherent in the dual nature of matter is clear from what has been said so far: we can know exactly the position of a micro-entity, but in this case, we sacrifice any determination of speed and vice versa, this, of course, applies to the results of observation acts, before any experiment, on the other hand, matter is always in a superposed state.

The principle of Complementarity[11] formulated at the end of 1927 by Bohr,[12] allows any contradiction to be overcome by eliminating

---

11   It should be made clear here that for Bohr the term complementarity is synonymous with complementarities. For Bohr, however, the word complementarity or complementarities is not to be understood as mutual functionality but as contradiction, exclusion.

12   One of the most lucid formulations of Complementarity given by Bohr is as follows: "Complementarity is the conception of objectivity in quantum theory.

any logical irreconcilability between the corpuscular and wave description (or between the "properties" of position and velocity, energy and time), implying the impossibility of demonstrating the truth of one or the falsity of the other.

Within the framework of the complementary model, the wave and corpuscle aspects manifested by the entities of the microworld come to be "complementary" elements of the same reality, come to be two sides of the same coin. Both are alternately necessary for a complete description of the observed reality.

In Complementarity, the contradiction due to the dualistic interpretation of matter becomes an "almost natural" impossibility of drawing a clear-cut line between the known object and the knowing subject: the manifestation of matter in a definite way in one form rather than another depends only on the type of experiment chosen to observe it.

In the course of his physico-philosophical research, Bohr aimed to elevate his idea related to the alternative and mutually exclusive character of microworld descriptions to the status of a general philosophical principle. For Bohr, ideas derived from complementarity relations can be applied to a multitude of situations, from biology to psychology. In his book Quanta and Life he writes in this regard:

> However unusual this development in physics may seem, I am sure that many will have recognised the close analogy between the situation in the analysis of atomic phenomena, as I have described it [in the Undetermined Principle], and certain characteristic aspects of the problem of observation in the field of psychology [...].In the course of introspection it is clearly impossible to distinguish clearly between phenomena per se and their conscious perception, and although it is sometimes permissible to speak of concentrated attention to any particular aspect of

---

Reality is never described in itself, but as a non-isolatable part of sets of phenomena and representations. The same reality can be the object of two complete, contradictory representations [...]. These two representations are called complementary. This is the case with the corpuscular and undulatory aspects of matter"...

a psychic experience, closer examination reveals that even in such cases one is in fact dealing with mutually exclusive situations. We are all familiar with the old adage that if we try to analyse our emotions, we cease to possess them; and it is in this sense that we have to be aware of the fact that our emotions are not the same as our own know that there is a complementary relationship between psychic experiences, with which we associate the words "thoughts" and "feelings"...

During a visit to the Far East, Bohr was profoundly impressed by the ancient Chinese idea of polar opposites represented by the symbol of *T'ai Chi*: the half-white, half-black circle that plastically depicts the archetypal opposites *yin* and *yang*.

When Bohr was awarded the noble Order of the Elephant in 1947 for his scientific merits, he chose the symbol of *T'ai Chi* (enclosed by the Latin motto "Contraria sunt complementa", i.e. "opposites are complementary") as his coat of arms (see figure below), which was to represent the complementary relationship between the opposite archetypes *yin* and *yang*.

Heraldic Coat of Arms of Niels Bohr

1.12. Energy-time uncertainty and its consequences

The energy-time uncertainty relation, with its consequences at the level of the energy conservation principle,[13] accounts for phenomena such as radioactivity or the emergence of *virtual particles*.

Suppose we want to determine the energy of a photon. According to the mathematical formalism devised by Planck, the energy (E) of a photon is directly proportional to the frequency (v) of light (in fact: $E = hv$). If, therefore, the frequency is doubled, the energy also doubles. A practical way of measuring energy is therefore to measure the frequency of the light wave, which can be done by counting the number of oscillations in a given time interval. In order to be able to apply this procedure, however, at least one and preferably more than one complete oscillation must occur, which requires a defined time interval. The wave must go from a maximum to a minimum, and then back to a maximum again. Measuring the frequency of light in less time than it takes for a complete oscillation is obviously impossible, even in principle. For light, the time required is very small (one millionth of a billionth of a second). Electro-magnetic waves with longer wavelengths and lower frequencies, such as radio waves, can take a few thousandths of a second to complete oscillation. These simple considerations highlight the existence of a fundamental limit in the precision with which the frequency of a photon (and hence its energy) can be measured in a given time interval. If the interval is less than a whole period of the wave, the energy is as indeterminate as ever. In order to have an exact determination of the energy, a "relatively long" measurement must be made (corresponding to at least one complete oscillation), but if what interests us is the instant at which an event

---

13   The principle of conservation of energy asserts that in nature "nothing is created and nothing is destroyed, but can only change state". In other words, it is impossible that in nature any form of energy (or mass) can arise from nothing, or that any form of energy (or mass) can be totally erased from the universe.

occurs, this can only be determined exactly at the expense of the energy information. We thus find itself having to choose between energy information and time information, which present an incompatibility similar to that for position and momentum.

From what has been explained so far, it is clear that the limits in the measurement of energy and time (as well as position and momentum) are not mere technological shortcomings, but objective properties by nature. In no sense can it be assumed that a photon (or electron, etc.) actually possesses a definite energy at a given instant. For photons, energy and time are two contrasting, "complementary"' characteristics; which of the two manifests itself more precisely depends only on the nature of the measurement one chooses (that the experimenter chooses) to make.

An immediate consequence of the above demonstrations is that for a very short time, the possibility opens up for a violation of the principle of conservation of energy.

For times of around a millionth of a billionth of a billionth of a second, an alpha particle (i.e. a "particle" formed by two protons and two neutrons) can borrow a certain amount of energy from the "quantum vacuum", cross the potential barrier that keeps it anchored to the nucleus, and escape from the atom to which it belongs. For times of around one billionth of a trillionth of a second, an electron and its antimatter companion – the positron – can suddenly form out of the vacuum, join and then vanish. All these phenomena are possible because the uncertainty relation $\Delta t \cdot \Delta E = h$, with its "inverse" $\Delta E = h/\Delta t$, makes it possible to obtain energy (or mass) from the "vacuum".

From the relation $\Delta E = h/\Delta t$, it is clear that as Δt *tends to* zero, ΔE tends to infinity. For very short times, therefore, it is possible to obtain considerable amounts of energy from the "vacuum"; this explains why high-speed alpha emissions (discovered in 1898 by Henri Becquerel) or the emergence of *virtual particles*.

1.13. Quantum Electrodynamics and vacuum fluctuations

An interesting consequence of the energy-time uncertainty is the fluctuation of the electromagnetic field. Quantum Electrodynamics (QED) deals with field fluctuations and merges two concepts: the electromagnetic field and photons as a corpuscular manifestation of the field itself. Through Quantum Electrodynamics, the phenomena of electric attraction and repulsion, magnetism, anomalies of the electron's magnetic moment, etc. have found consistent explanation.

The quantum field is a completely new concept that has been applied to the description of all subatomic particles and their interactions, with each type of particle corresponding to a different type of field. The quantum field is seen as the fundamental physical entity: a continuous medium present everywhere in space.

In Quantum Electrodynamics, the interaction between two microparticles (two electrons, for example) occurs through the exchange of virtual photons. The virtual photons are born and live thanks to the very small degree of uncertainty existing between the energy and time levels, which is well highlighted in the previous Section and which characterises the "objective state" of entities and events in the microworld.

This new conception of subatomic interaction may seem difficult to understand and represent, but it becomes relatively easy and accessible when the process of exchanging a virtual photon is depicted by means of a Feynman space-time diagram[14] Figure 2 shows a Feynman diagram in which two electrons are represented approaching each other; one of them emits a virtual photon (denoted by γ) at point A, the other absorbs it at point B. After emitting the photon, the first electron deviates from its direction, and so does the second electron when it absorbs the photon.

---

14   Feynman diagrams are the simplest and most effective tool for representing electromagnetic interactions. In 1965, Richard Feynman was awarded the Nobel Prize in Physics for his contributions to the formulation of Quantum Electrodynamics.

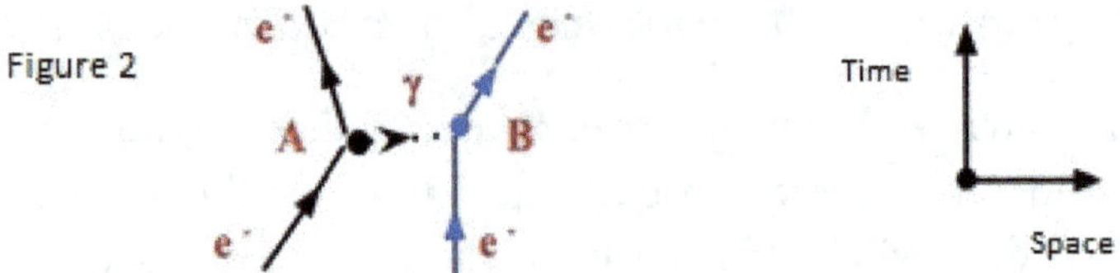

The complete interaction between electrons occurs through a continuous series of photon exchanges.

Every electrically charged particle continuously emits and reabsorbs virtual photons and/or exchanges them with other charged particles. When two electrons (i.e. two charges of the same sign) exchange virtual photons, they repel each other; when a proton and an electron (i.e. two charges of opposite sign) exchange virtual photons, they attract each other. In terms of classical physics, one could say that the particles exert a repulsive or attractive force on each other.

It is interesting to note that in the above-mentioned "force exchanges", neither particle actually "collide" with the other: they simply interact (*inter-act*) by exchanging photons, and the force is nothing more than the collective macroscopic effect of these repeated photon exchanges. The concept of force in subatomic physics must therefore give way to the concept of interaction (*inter-action*) between particles through fields consisting of sets of virtual photons that live the time necessary to fulfil their task. In the case of electromagnetic interactions, the virtual field particles exchanged are photons; in the case of interactions between the components of the atomic nucleus, the particles exchanged are mesons.

The "inter-action" mechanism just described can be further exemplified by imagining two skaters standing still on a sheet of ice a few metres apart. The repulsive force between them can be illustrated by exchanging an object: a heavy ball, for example. The first skater throws the ball and consequently repels; the other receives the ball and is pushed "sideways" by it, so that at the end of the process, the two skaters move away from each other with diverging trajectories. The two skaters in the example represent the particles interacting, the ball the virtual particle.

When, as in Figure 2, an electron emits a virtual photon that is absorbed by another electron, the microparticle is said to be interacting with another particle. When, on the other hand, an electron emits a photon and re-absorbs it, it is said to be interacting with itself. Self-interaction makes the world of subatomic particles a kaleidoscopic reality of incessant transformation processes. Protons, like electrons, can interact with themselves in more than one way. The simplest self-interaction that a proton can perform is the emission and reabsorption of a virtual particle called a neutral pion (see Figure 3) within the time frame allowed by the Principle of Uncertainty.

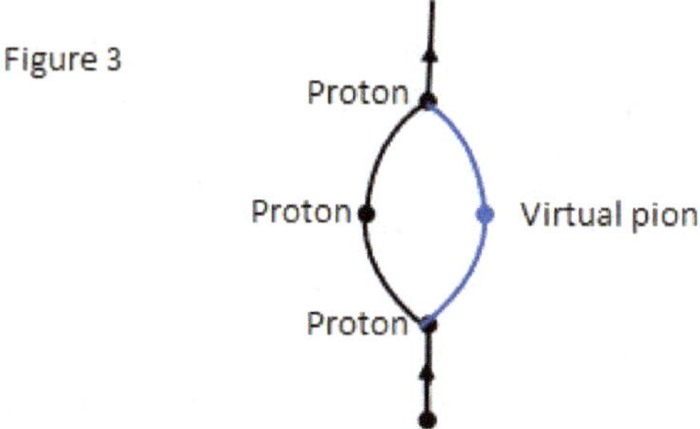

If, however, for whatever reason, the proton (or any other "mother" particle) were to disappear, the virtual pion could no longer be reabsorbed; it would then assume the status of a real particle itself. This is what happens when a proton meets an antiproton, both of which suddenly disappear, leaving the pion, or a cloud of virtual photons. These virtual particles thus present themselves to the world as real particles, since their "debt" to the Principle of Uncertainty has been settled by the mass-energy of the vanished proton-antiproton pair. Figures 2 and 3 illustrate the creation (and exchange) of virtual particles in the presence of real entities: electrons and protons. The formalism of Quantum Electrodynamics predicts that a process of virtual particle creation can occur even in the absence of matter. The diagram in Figure 4 shows a process of creation and destruction (annihilation) in the quantum vacuum of virtual particles of opposite charge electron-positron (the positron is a positively charged electron).

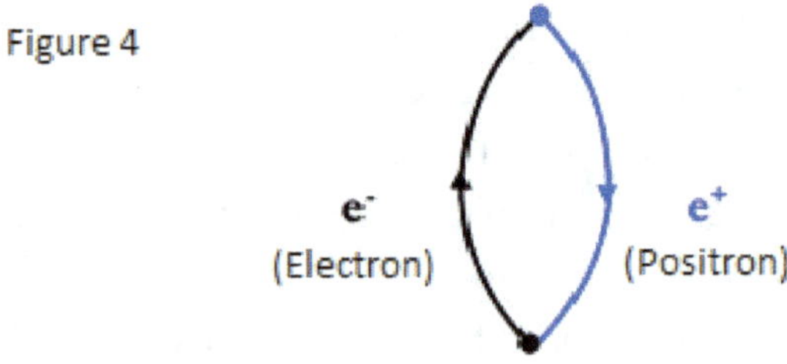

In Quantum Electrodynamics, the vacuum undergoes a kind of "polarisation" (as a result of spontaneous fluctuations in the electromagnetic field), allowing matter to come into being spontaneously. The average lifetime of the virtual pairs created is, however, very short. For the electron-positron pair, for example, it cannot exceed one billionth of a trillionth of a second (the time allowed by the Uncertainty Principle to have mass/energy from the "vacuum").

The subatomic events just described have given physicists a new perspective for understanding empty space. For Heinz Pagels of Rockefeller University, the vacuum resembles the surface of the ocean:

> Imagine flying over the ocean in a jet. From that optimal vantage point, the surface looks perfectly even and flat. But you know that if you were on a boat, you would see huge waves all around you. This is how the vacuum behaves. Over great distances – the distances we experience as human beings – space appears completely empty to us. But if we could analyse it very closely, we would see all the quantum particles going in and out of the vacuum.

The concept seems to defy common sense but is perfectly valid in the context of Quantum Electrodynamics: "There is no more fundamental point than this," wrote physicist John Wheeler, and again "Empty space is not empty. It is actually the region where the most violent physical phenomena occur".

1.14. The "Measurement Problem" in quantum mechanics

One of the fundamental problems encountered in quantum mechanics is basically related to measurement processes. Any attempt to extract information on the state of a physical system (as we have seen in the uncertainty relations) encounters difficulties mainly due to the intervention of the instrumental apparatuses used, which significantly alter the observed physical system.

If, in classical physics, the problem of the influence of the instrument on the quantity being measured does not generally arise (e.g. the observation of the position of the moon through a telescope does not produce any appreciable alteration in the state of this body, and the instrument can be considered to be external to the theoretical elaborations in which it is used), on the other hand, with electromagnetism, the concept of "ideal measurement" already began to show, so to speak, the first signs of crisis, since the measurement of fundamental electrical quantities, which is carried out by means of ordinary electromagnetic instruments, disrupts the physical *status of* the "entity" being measured.

However, it is with quantum physics that the measurement problem arises in full force. This is due to the fact that, in non-quantum physics, the perturbation induced by the instrument can, at least in principle, be made arbitrarily small until it becomes practically irrelevant, by means of the construction of sophisticated and adequate apparatuses (think, for example, of all those instruments based on the principles characterising Kelvin's absolute electrometers). The existence of Planck's quantum of action, on the other hand, introduces in the physics dealing with elementary particles a totally new point of view regarding the description of the measuring process: every interaction between a micro-object M and an instrument D is no longer reducible to a recording by D of the value of the quantity measured on M, but implies an important modification also of the system M. This modification cannot be eliminated either practically or conceptually. "There exists," writes P.A.M. Dirac, "a limit to the degree of fineness of our means of observation, and consequently a lower extreme for the magnitude of the perturbation accompanying the observation itself, a limit that is inherent in the very nature of things and cannot be overcome by better techniques or greater skill on the part of the observer".

In fact, the problem arises from the fact that in order to acquire information about the properties of a micro-object, one cannot but perturb it, and this perturbation cannot be made arbitrarily small, but must have Planck's constant as its minimum value. In practice, as the Uncertainty Principle indicates, any interaction between measuring equipment and observed elementary particle must always involve the transfer of a certain momentum over a certain spatial distance or energy exchange for a certain time.

Now the physical dimensions of "energy for time" and "momentum for space" are precisely those of action, that is, they are those of Planck's constant (h). The perturbative character of measurement in microphysics (Tarozzi makes this clear in his writings today) forces us to abandon the classical scheme of registration in favour of the following scheme:

$$M \leftrightarrow D \to O$$

which clearly indicates the symmetrical and no longer unidirectional nature of the interaction process between measuring apparatus D and micro-object M. Now applying Schrödinger's wave equation Ψ (which, as we have seen in the previous pages, describes the state and evolution of systems in quantum mechanics) to the description of the measurement process just represented, that is, interpreting this process as any process of interaction between two physical objects (corresponding to the micro-object M and the measuring apparatus D), every time M is in a state of superposition, the final state of the overall system M + D is, due to the *linearity of* the Schrödinger wave equation, also in a state of superposition, in this case of different macroscopic states.

Such a possibility would imply that observer O, who controls apparatus D, would notice the latter simultaneously recording all the different results predicted by the quantum mathematical formalism (in practice, D would also enter a state of superposition), which, apart from being paradoxical, makes any act of measurement theoretically impossible. Practically speaking, according to the formalism of quantum mechanics (expressed in our case by the Schrödinger equation), no measurement would be possible, as all the entities involved in the measurement chain (including the observer's sense organs) would enter a superposed state, effectively preventing any measurement.

In order to avoid this contradictory conclusion, various solutions have been proposed, depending on the answer to be given to the level at which the process of reduction of the wave function should take place; that is, the transition from a superposition of different physical states to a definite final state, to a definite measurement. This is, of course, a problem of considerable relevance to the phenomena of physics since it directly affects two closely related fundamental questions: the domain of application of the quantum formalism and the ability to describe the measuring instrument by the formalism itself.

Bohr's permanent trace

The first answer to the measurement problem comes from Bohr's "Measurement Theory", which states that the reduction of the wave function (and thus the measurement) occurs at the level of the measuring instrument. This hypothesis thus recognises a special status for the instrument compared to ordinary macroscopic systems, whose interactions with micro-objects continue to be described by the Schrödinger equation. In Bohr's case, the scheme of the measurement process is of the type $M \leftrightarrow D \rightarrow O$ and the validity domain of the quantum formalism extends to microscopic and macroscopic systems, excluding measuring instruments; which would never enter a superposed state.

The Subjectivist Solution of von Neumann and Wigner

The Second Answer to the measurement problem is contained in von Neumann's "subjectivist theory". It assumes that the reduction process of the wave function occurs at the level of the consciousness of the observer. This answer allows the instrument to be included among the objects describable by the quantum formalism, whose domain of validity is thus extended to all physical objects, including measuring instruments of course, excluding the consciousness of the observer. This, conceived as an agent external to the physical world, is responsible for the reduction of the wave function, which occurs within a rigorously symmetrical process, corresponding to: $M \leftrightarrow D \leftrightarrow O$

Basically, what happens is that the observed particle "pours" its superposition onto the measuring apparatus, the measuring apparatus onto the sense organs and these onto the observer's brain. The chain of "reciprocal conditionings" could never end if it were not for the observer's consciousness (or "faculty of introspection"), which, due to its *irreducible* characteristics, allows the chain of superpositions to be severed, thus enabling the "measurement" of the micro-object, allowing reality to take on consistency.

Von Neumann's theory was taken up in recent times by Nobel Prize winner Eugene Wigner, who formulated a version of it called "interactionist". According to this theory, there are two levels of reality. A primary (and fundamental) reality that is mental reality. A secondary (and relative) reality that includes all other "things" (matter, space, etc.). According to Wigner, the act of measurement by a "conscious being" interacting with the material world modifies the physical world through the reduction of the wave function, enabling the measurement process.

The Many Worlds Theory

The Third Answer to the measurement problem is contained in the "Many Worlds" theory formulated in 1957 by Hugh Everett III (later taken up and expanded by Bryce De Witt). It states that the reduction of the wave function never actually takes place and that all the possibilities envisaged by the formalism are realised simultaneously in different branches of the universe. This restores reality to the wave function by considering it a true description of the universe. The reward for this promotion consists precisely in the overcoming of the measurement paradox, for there is no need for a special reduction to reality at the moment of observation.

In the Many Worlds theory, every time a measurement of a micro-object is made, the universe splits into two physically real and "identical" universes. The only difference between these universes lies in the different "state" of the observed particle. In the case of measuring the spin of an electron,[15] for example, in one universe the spin would be UP, in the other universe it would be DOWN, and the universes would otherwise be the same.

In Everett's perspective, the domain of application of quantum theory extends to macroscopic objects, microscopic objects M, measuring instruments D and observers O themselves, all conceived as

---

15   Spin is the momentum of rotation of a particle about its axis. If, for example, the rotation of an electron is "right-handed", spin is said to be UP; if it is "left-handed", spin is said to be DOWN. For quantum mechanics, however, before a measurement an electron possesses, superimposed, the two states, "spin UP" and "spin DOWN", and describes them using a single wave function.

elements of physical reality, while the scheme of the measurement process, which considers the branching process of the universe, can be exemplified as follows:

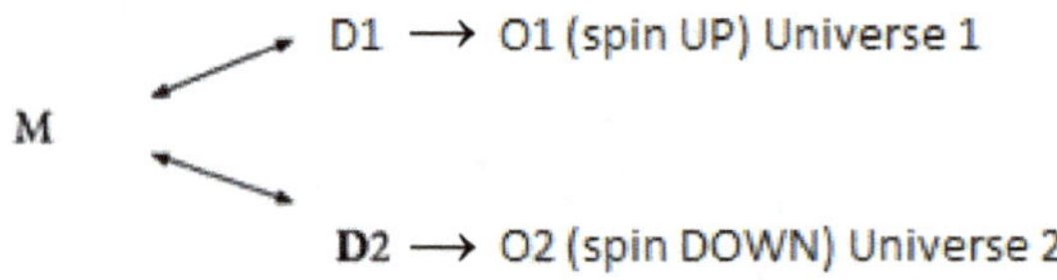

Interestingly, in some passages of the short story *The Garden of Forking Paths* written in 1941 by Jorge Luis Borges, we find singular points of contact with the "Many Worlds Solution".

For example, Borges writes:

> A new reading of the whole work confirmed this idea for me. In all narrative works, once one is confronted with several alternatives, one decides – simultaneously for all of them – [...] Thus, several futures, several times are created, which in turn proliferate and bifurcate [...] This plot of times that come together, bifurcate, cut off or ignore each other for centuries encompasses all possibilities.

Theory of many stories

An alternative to the physical division of universes was proposed by Nobel Prize winner Murray Gell-Mann and James B. Hartle based on an idea by R. B. Griffiths. This alternative – called the "Many Stories" alternative – envisages that a set of mathematical prescriptions would describe the probabilities of all possible "stories" that can be imagined characterising the unfolding of a measurement process. Such stories would however only describe "potentialities" and not physical realities as in the Many-Worlds solution.

Theory of Many Minds

A second alternative to the physical division of universes has been proposed by David Albert and Barry Loewer. In their theory – known as the "Many Minds" theory – the two American scholars hypothesise that each "sentient" observer or physical system is associated with an infinite set of mental states that experience the different possible outcomes of quantum measurements. The family of latent choices in the Schrödinger equation would thus correspond to the myriad of experiences undergone by these minds and not to an infinite number of material parallel universes.

The "dynamic reduction" and macro-reduction theories

The Fourth Response to the Measurement Problem groups together the theories that predict that the reduction of the wave function occurs through a "dynamic process", i.e. through the transition from the microscopic to the macroscopic level. The various attempts that have been made in these directions are listed below:

a. Landau's Superselection Rules, which *prohibit the overlapping* of macroscopic states.

b. the measurement theories of Daneri-Loinger-Prosperi, Ludwig, Prigogine (and co-workers). According to these scholars, the scheme of the measurement process is similar to Bohr's, but the domain of application of the quantum formalism is restricted to the micro-objects, assuming that the description of macroscopic phenomena, including that of interaction, requires the elaboration of a *new quantum macrodynamics*, in the absence of which it is appropriate to resort (for the description of the measurement process) to classical or semi-classical theories such as stochastic mechanics.

c. The Ghirardi-Rimini-Weber (GRW) theory, which states that the reduction of the superposition of states *occurs spontaneously* as the microscopic dimensions of the particle are shifted to the macroscopic dimensions of the measuring apparatus. According to this theory, as time passes, the on-function of a particle "expands". During the expansion, there is the probability that the wave hits "something in the background" and suddenly becomes localised. For single particles, the probability of such a collision is very low, about 1 in 100 million years, but at the size of the measuring instruments, the probability of the reduction of the wave function rises to 1 in 100 picoseconds (1 picosecond = 10^{-12} seconds).

d. Very similar to GRW's theory is the theory based on "spontaneous decoherence" developed by Wojciech Zurek. It is based on the idea that "the environment" destroys the quantum superposition state.

e. Penrose's theory, finally, states that the reduction occurs by *spontaneous collection* of two different states of the space-time metric as a result of the action of Quantum Gravity (at Planck dimensions gravity becomes Quantum Gravity). Now, in line with the postulate of the superposition of the states of quantum mechanics, there would be two different superposed gravitational fields; there will be a time when the space-time geometries of these two fields are sufficiently different from each other that Nature is forced to "choose" between them, operating some kind of wave reduction. The greater the energy difference between the superposed states, the shorter the duration of the superposition: for a proton the reduction time is of the order of millions of years; for a mass equivalent to a grain of dust it is of the order of a millionth of a second.

Conclusions

Let us now see what the limitations and strengths of the four indicated ways are to solve the Measurement Problem.

1. The solution indicated by Bohr has the major limitation of renouncing the elaboration of a rigorous mathematical theory for the description of interaction processes, as well as the assumption of a principled difference between measuring instruments and any other macroscopic system.

2. The subjectivist solution of von Neumann and Wigner, while boasting a remarkable logical consistency and describing measuring instruments like any other physical object, has the limitation of introducing what could be described as "idealistic" elements into physics, leaving, among other things, the great doubt as to what the term *"conscious observer"* actually means.

3. The "Many Worlds" solution, while possessing a relevant formal consistence and describing measuring instruments like any other physical object, has a blatantly paradoxical character. Each measurement process would produce almost identical copies of physically real universes that then branch out indefinitely. Moreover, every time a measurement process takes place, all the possibilities provided by the formalism are realised in different worlds that are inaccessible to each other. If the worlds are all inaccessible to one another, there is no empirical possibility of verifying the rightness or wrongness of the "Many Worlds" solution.

4. The "dynamical reduction" and macro-reduction solution, while currently enjoying the greatest consensus in the physics community, has the major limitation of adding a non-linear term to the wave function (which, as seen above, is described by a *linear* equation), so that when the dimensions of the system tend to become macroscopic, its superposed states converge into a single state.

1.15. Negative Outcome Measurements and Subjectivism

In relation to the "Dynamic" and Macro-realistic theories proposed up to now to solve the Measurement Problem, there is a serious objection put forward by Wigner, which makes use of a paradox proposed by Renninger.[16]

This objection highlights the existence of measurement processes which, although correctly informing the observer about the state of the measured particle through the reduction of the wave function, occur without any controllable modification of the measuring apparatus.

Wigner's argument, which as mentioned takes its starting point from Renninger's paradox, is based on the following Gedankenexperiment (i.e. an imaginary situation so convincing that it does not require being reproduced in practice).

Let us consider (see figure on next page) a source of a single photon E (the speed of the photons is 300,000 kilometres per second) that isotropically emits these particles in all directions and is partially surrounded by a hemispherical screen (Screen 1) of centre E and radius R_1 (R_1 = 600,000 kilometres) and a second spherical screen (Screen 2) of radius R_2 (R_2 = 1,000,000 kilometres).

---

16   By means of a conceptual experiment, Renninger writes, it is shown that, contrary to current opinion, there are measurement processes that exert no action on the measurement object.

These negative measurements consist of the experimental determination of the as- semblance of events that were expected with a certain probability, determinations that – the mark of a true measurement – give new predictions about the measurement object; thus, they cause a reduction in the wave function in exactly the same way as normal, positive observations that perturb the measurement object. (Renninger, 1960).

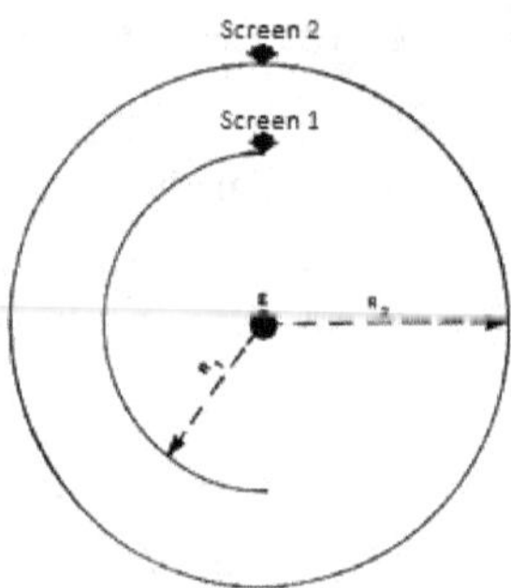

Assume further that both screens are covered with a sensitive substance capable of detecting photons. Thus, the photo emitted by source E can be absorbed by either Screen 1 or Screen 2. Now the probability that a photon will strike either Screen 1 or Screen 2 after a certain period since emission is given by $\Psi = 0.5\,\Psi_1$ (Screen 1) $+ 0.5\,\Psi_2$ (Screen 2).

We will then have two possibilities:

1. The first possibility is the detection of the photon on the first screen, which implies the cancellation of Ψ_2 and the transformation of Ψ_1 into certainness. According to the postulate of the reduction of the wave function, the new physical situation will then be described by: $\Psi = \Psi_1$ (Screen 1). In this case, the reduction process can be explained on the basis of the macro realistic and "dynamic" measurement theories, as a consequence of the interaction between the revealed photon and the measurement apparatus (consisting of Screen 1), which as a result of this interaction will evolve towards the stable state of thermo-dynamic equilibrium, given by a well-defined state vector Ψ_1 (Screen 1).

2. The second possibility is that after two seconds the detection of the photon by Screen 1 does not take place. This implies that due to the postulate of the reduction of the wave function, a fraction of a second after the non-detection of the photon by Screen 1, we are certain, without the need for any measurements, that the photon will strike the second screen: the physical situation is then described by $\Psi = \Psi_2$ (Screen 2).

In the latter case, however, the change in mathematical description cannot in any way be related to any physical process of interaction between the photon and the macroscopic instrument (in our case Screen 2) but turns out to be a consequence of the information that we, as (conscious) human observers, have deduced/obtained from a certain physical situation, i.e. by not recording the occurrence of a given phenomenon. This means that to explain the transition from $\Psi = 0.5\,\Psi_1$ (Screen 1) $+\ 0.5\,\Psi_2$ (Screen 2) to $\Psi = \Psi_2$ (Screen 2), we have not resorted to a theory of measurement that regards the instrument as a macroscopic entity in a metastable state that evolves towards stability as a consequence of interaction with the measured object, but to the simple acknowledgement by the conscious observer of a fact that did not occur; i.e. the non-disclosure by the photon on Screen 1.

Renninger's paradox seems to bring an apparently decisive argument in favour of the subjectivistic von Neumann-Wigner theory, which appears to be the only one capable of explaining how the acknowledgement by a conscious observer of the absence of a process of interaction between instrument (Screen 1) and measured object can equally lead to the reduction of the wave function.

Consciousness would thus come to play a fundamental role in the measurement processes in physics with considerable implications in terms of gnoseology, philosophy and the understanding of reality.

1.16. The "Schrödinger's cat" paradox

In 1935, Erwin Schrödinger, in an attempt to demonstrate the incompleteness and contradictions inherent in quantum theory, proposed the following thought experiment in a paper that has now gone down in scientific history (Schrödinger, 1935):

A cat is placed inside a steel chamber together with the following contraption [...] in a Geiger counter there is a small quantity of a radioactive substance, so that perhaps in the space of an hour one of the atoms will decay, but also, with equal probability, none will undergo this process. If it happens that one of the atoms decays, the counter generates a discharge and through a relay releases a hammer that shatters a small glass vessel containing prussic acid. If the whole system has been isolated for an hour, it can be said that the cat is still alive if no atom has undergone a decay process in the meantime [...].

The wave function of the system will express the complete situation by means of the combination of two terms referring to the living cat and the dead cat (pardon the expression), two situations mixed or blended in equal parts.

According to Schrödinger, therefore, insisting on the completeness of the quantum theory means admitting that one hour after the start of the experiment, the wave function of the system, i.e. the wave function of the radioactive substance (the state of the radioactive substance) to which the life of the cat (the cat's state) is connected, will be represented by the linear superposition of two terms:

$$\Psi(\text{system}) = \Psi 1_{(\text{no atom has disintegrated; the cat is ALIVE})} + \Psi 2_{(\text{one atom has disintegrated; the cat is dead})}$$

Furthermore, continues Schrödinger, "if one wants to follow the postulates of quantum theory to the letter, if after an hour from the start of the experiment the chamber is opened and it is found, for example, that the cat is dead, it must be admitted that it was the act of observing that killed it by eliminating the part representing the living cat from the superposition".

It should be pointed out here that the + sign of the formula $\Psi_{(cat)} = \Psi1_{(live\ cat)} + \Psi2_{(dead\ cat)}$ implies that it cannot be assumed that the cat is either alive or dead, and that this co-presence of incompatible potentialities can be eliminated only by making an observation to determine whether the cat is alive or dead. In a literal sense, if we recognise the fact that looking into the box, we certainly find either the cat alive or the cat dead, we are forced to conclude that it is our act of co-scientific observation/perception that determines the cat's benign or lethal fate.

Also in the context of the analysis of the consequences of "observational acts".

A further paradoxical aspect emerges. If the experimenter decides to postpone observation of the chamber indefinitely, the cat remains in its schizophrenic state of latent life until it is given a definitive dimension, by virtue of the experimenter's polite but capricious curiosity.

Some scholars have argued that a cat cannot be considered an appropriate observer as it does not possess the full awareness of its own existence that humans do, so it would be too "inarticulate" to know whether it is alive, dead or "alive-dead".

To circumvent this objection, it was suggested (again in a thought experiment) to replace the cat with a human being. Generally in the physics community, this volunteer is known as "Wigner's friend", as it was Eugene Wigner himself who analysed this aspect of the paradox (Wigner, 1961).

1.17. The "Wigner's friend" paradox

Such a worthy accomplice installed inside the steel chamber could now be asked, in the event of his survival in the above experiment, how he felt during the experiment itself. No doubt he would answer "fine", despite our assumption that his body was in a "life-and-death" condition for the duration of the test.

In reality, the real problem is to establish the way in which Wigner's friend responds in a certain way to the question formulated by the person (Wigner) who is (during the experiment) outside the chamber, as well as to understand at what level the reduction of the state of superposition takes place. It is clear that the paradox of Wigner's friend confronts us with a contradictory situation in which two beings with "consciousness" (one of whom is inside the steel chamber) are involved in the same experimental situation and "compare" the results of their observations. Who, in this view, is really responsible for the reduction of the of the overlapping state?

Following a logic close to orthodox quantum theory, only the observant outside the chamber, is able to make this reduction. In fact, he who is inside the chamber (Wigner's friend), being "part" of the experiment, of the instrumentation, must inevitably be described by the same wave function that describes all the objects inside the chamber (which one hour after the start of the experiment are in a combination of two states), which makes him unsuitable for reducing the superposition state. Only those outside the "steel chamber" system (i.e. Wigner) can reduce the superposition state by their act of observation.

At this point and from the perspective just described, it is evident that quantum mechanics becomes a theory that leads to the most exasperated solipsism. Only a particular observer in fact (in our case Wigner and not his friend or another living being), is able to produce and know true reality; a reality different from that experienced by other human beings. In the past, the attitude whereby only the subject and his perceptions are real, while all other things and people are "superstructures", had reaped plaudits first and foremost among Cartesians; Wigner, however, finds it unacceptable.

To escape solipsism, Wigner advances the hypothesis that quantum mechanics does not apply to human beings. How this conclusion can be reconciled with the fact that quantum theory must describe material reality in all its aspects has already been seen in von Neumann and Wigner's solution to the Measurement Problem: "The consciousness that characterises human beings is an irreducible attribute that breaks the chain of superpositions by producing the reduction of the wave function".

According to Wigner, therefore, the friend inside the chamber will never fall prey to superposition states. He, in fact, despite being in the same condition as the cat, will, like any other being endowed with consciousness,[17] be able to immediately and autonomously reduce the state of superposition, understanding, among other things, "the turn" that his immediate destiny is taking.

It is not surprising that Wigner's ideas have attracted the widest criticism. Scientists normally look at consciousness as something that, at best, is ill-defined and, at worst, does not physically exist.

However, it must be considered that all our observations, and through them all science, are ultimately based on our awareness of the world around us.

According to the usual conception, consciousness can be estimated by the external world but without in turn influencing it, thus violating a Principle – the Principle of Symmetry – that is in some respects universal.

Wigner therefore proposes not to disregard the Principle of Symmetry in the case of "mind-world" interactions, proposing an action of awareness on matter consisting of the possibility of reduction by the mental substratum of the superposition state of matter.

---

17   "My friend," writes Wigner in *Two Kinds of Reality*, Indiana University Press, Bloomington (1967), "has the same kinds of impressions and feelings as I do…"

A decidedly special aspect arising from the acceptance of Wigner's conception occurs in the case of two "conscious beings" observing the same system at different times and in different ways.

To illustrate the problems that may arise in such a case, let us consider, as in the previous situations, an atom whose decay activates a Geiger counter, but this time in the absence of any observer immediately involved in the experiment and where the counter (via the relay) commands the displacement of an index.

Suppose now that after one minute, when the probability of the atom's decay is 50%, the experiment ends, and the counter pointer locks into its current position: if the atom has disintegrated, the pointer will be displaced, if not, it will be in the clockwise position. The index can then be read at any later time. Now, instead of having an experimenter "observe" the index directly, the Geiger counter is photographed. Once the photograph has been developed, the experimenter looks at it, without ever consulting the counter directly. According to Wigner, it is only at this final stage of the procedure that reality takes on substance; because reality owes its "actualisation" to the conscious act of observation performed by the experimenter or anyone else. One should therefore conclude that, before photography was examined, atom, Geiger counter and photograph were all in the schizophrenic states due to the coexistence of superimposed states. All this, even if the development of photography were to be postponed for several years. This little corner of the universe would then remain suspended in the unreality until the experimenter (or some curious passer-by) deigns to take a look at the photograph.

The truly paradoxical problem arises when, at the end of the experiment, we take two successive photographs of the index connected to the meter; photographs which we shall call A and B. Since the index has been locked, we know that picture A must be identical to picture B. The trouble begins when two experimenters – let's call them Anna and Bruno – enter the scene, and Bruno looks at picture

B before Anna looks at picture A. Thus B was taken after A but was looked at before. Wigner's theory requires that Bruno is in this case the conscious individual responsible for the creation of reality, and this is because he looks at his photographic record first. Suppose Bruno sees the shifted index finger and therefore states that the atom has disintegrated. Naturally, when Anna looks at photograph A, it too will show the shifted index finger. The problem lies in the fact that when photograph A was taken, photograph B did not yet exist, so in some mysterious way the act of looking at B performed by Bruno is the cause of A becoming identical to B, even though A was taken *before* B. One therefore seems to be obliged to believe in a kind of *retroactive causality*. In the next section, we will see a typical situation of *retroactive causality* realised in an experimental context.

1.18. Delayed choice experiments

At the end of the 1970s, one of the most prestigious American physicists, John A. Wheeler (Bohr's pupil and Feynman's teacher), using an instrument called the Mach-Zehnder interferometer (this instrument closely resembles the two-slit apparatus seen in one of the previous Sections) demonstrated that there are situations in which it is possible to witness a reversal of the time order of phenomena. Before explaining in detail what the experiments proposed and performed by Wheeler and co-workers (known to the scientific community as "delayed-choice experiments") consist of, it is worth making a few clarifications on how Mach-Zehnder interferometers work.

Figure 5 shows what happens to a photon (or electron, or any micro entity) entering a Mach-Zehnder interferometer. The first thing the photon encounters is a semi-transparent mirror (M): the reader will certainly have seen a version of said mirror in some detective films, in situations where it was necessary to see without being seen.

Following interaction with M, the photon has a 50% chance of either attracting it (thus heading towards mirror B) or being deflected (heading towards mirror A). Regardless of the branch taken and the deflection undergone, the photon will finish its journey on screen S. In the absence of S, the photon will impinge on detector P_1 or P_2. Detectors are devices that permanently record the arrival of particles such as photons, electrons, etc.

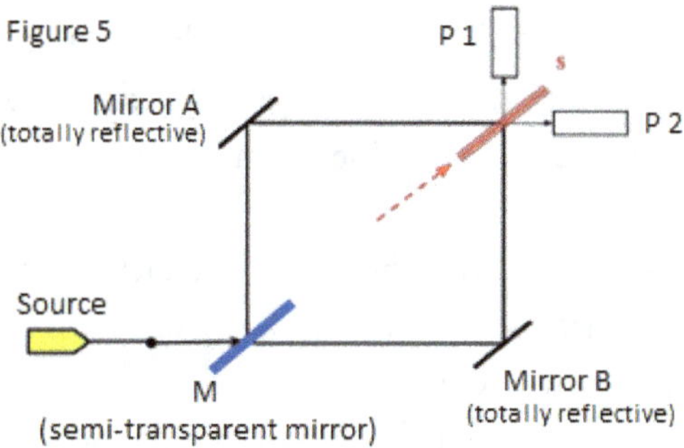

Let us now imagine a situation in which two photons cross the two branches of the interferometer (i.e. the branch with mirror A and the branch with mirror B) at the same time. Given the wave component associated with the photons, it will happen that the encounter of the photons in S will produce a typical interference situation. At this point the laws of optics (and common sense) tell us that such interference can only occur if both branches of the interferometer are crossed by a photon. In the case where only one photon passes through the interferometer, having only one wavefront no interference in S can occur. Instead, as in the two-slit experiments, even if only one photon is sent at a time, interference is produced on the S-screen. The quantum explanation of this phenomenon (as in the two-slit tests) is that the photon, after passing through the mirror M, "splits in two", travels through both branches of the interferometer at the same time, and in S self-interacts with itself generating constructive or destructive interference. This situation, although extremely simplified (for various reasons the delays that a photon accumulates while interacting with the mirrors were not taken into account), is the one produced by performing experiments with a single photon moving inside a Mach-Zehnder interference meter.

Let us now return to Wheeler and his tests. The experiments proposed and conducted by the American physicist focus on the possibility that screen S can be inserted immediately after the photon has interacted with mirror M (i.e. while the photon is in flight between M and S). By performing this operation, the facts show that the photon produces the light and dark bands typical of specifically wave-like behaviour.

Outlined:

a. If we insert S after the photon has interacted with M, we will have undulatory behaviour: on S, we will have destructive or constructive interference.

b. By not placing S after the photon has interacted with M, we have corpuscular behaviour: the photon is registered by either the P1 detector or the P2 detector.

At this point, however, Wheeler points out that something very strange has occurred. In fact, the wave or corpuscular reality must be assumed by the photon (as well as by any other micro entity) not at the level of S, P1 or P2 , but at the moment it interacts with the mirror M.

It is at the level of the semi-transparent mirror M that the photon is "forced" to manifest itself as a wave entity, thus passing through both branches of the interferometer, or as a corpuscular entity, passing through only one of the branches. In our case, however, the photon, which has interacted with M and with the screen S not in position (i.e. not in front of the detectors), must arrive as a corpuscular entity at the end of its course in the vicinity of S. How then can the screen S (inserted at the last moment) record the photon as a wave if after the interaction with the semi-transparent mirror M the photon had assumed the characteristics of a corpuscle, passing through only one of the interferometer branches? Wheeler's explanation of these facts is that whether or not the S screen is inserted immediately after the photon has interacted with M produces an effect in the past, "forcing" the photon to change its state. Basically, the choice (in the future) to insert S, conditions the way the photon propagates (in the past).

To better understand what has just been illustrated, let us look at what Wheeler himself writes about the meaning of delayed choice experiments: "Recording instruments operating in the here and now have an undeniable role in generating what has happened [...]. Quantum Physics shows that what the observer does in the future defines what happens in the past". And again in the paper entitled "Delayed Choice Experiments and the Bohr-Einstein Dialogue", delivered in London (1980) at the joint meeting of the American Philosophy and the English Royal Society, Wheeler states:

> It is wrong to think of the past as already existing in every detail, the past is theory. The past has no existence except for being recorded in the present [...]. What we have the right to say about past space, and about past events, is decided by choices – of what measures to take – made in the recent past and in the present. The phenomena brought into existence by these decisions extend backwards in time in their consequences [...]. Recording instruments operating here and now have an undeniable role in generating what appears to have happened. However useful it may be in everyday life to say "the world exists out there independently of us", this point of view can no longer be maintained. There is a strange sense in which ours is a participatory universe.

1.19. Is reality such and not otherwise because that is how we construct it?

What can be said, for instance, about the global reality, the reality of the universe? Applying quantum mechanics to the universe, some scientists have speculated that, like virtual particles, our universe may have arisen from a gigantic quantum fluctuation (we examined and will examine this in more detail in the previous Sections and later in Chapter Two), and like virtual particles that live for a very short time, our universe too may have been in a "non-real", temporally "relatively short", virtual phase of life.

What would "save" our universe from disappearance?

An "act of observation" by a conscious being, say some scientists. The action of consciousness or awareness in certain situations related to quantum mechanics plays a fundamental role in the creation of reality itself; we have already seen this, for instance, in the Subjectivist Solution to the Measurement Problem, or in experiments with negative results, or even in Schrödinger's Cat paradox.

Almost all scientists are reluctant to hypothesise a "consciousness" that beyond matter determines and gives substance to the universe, but there is a group of scholars who, following John Wheeler, believe that the current consciousness (ours or that of other "inhabitants" of the universe) has determined and now determines the universe as it is. Practically speaking, the Big Bang, the expansion, the formation of galaxies, the birth of stars, the appearance of life on certain planets, and its evolution into intelligent forms, all of this until the appearance of "conscious" life forms, would have been only a "virtual history": in reality, the universe would have been in an indifferent and essentially unknowable form (something reminiscent of the Kaos of Greek mythology) until the appearance of consciousness determined it as it is.

Consciousness, therefore, would have imposed a Cosmos on a formless Kaos (and thus also time, space, evolution and the appearance of consciousness itself). Ultimately, the universe produced consciousness, which in turn made the universe real through an act of observation!

At a major cosmology conference years ago, a drawing appeared (see figure below) that expresses this paradox well.

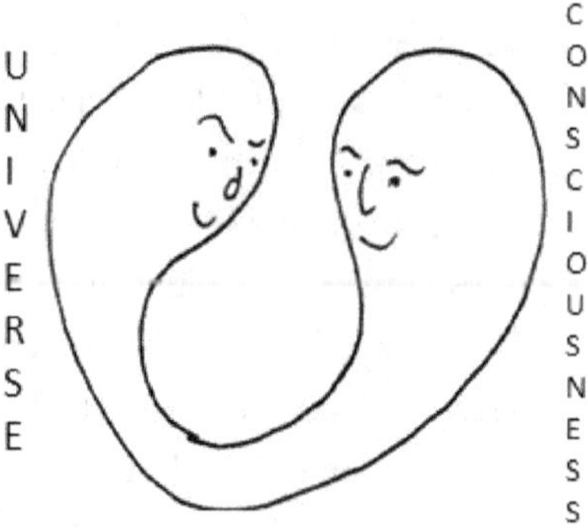

Although the idea seems absurd and the rational mind recoils at the thought that the universe may have existed for billions of years of "virtual life" before the appearance of consciousness, this interpretation held by some cosmologists may be more than fanciful speculation. In fact, the "delayed choice" experiments (we have seen such experiments in the previous Section), as well as all the tests that call into question the role of consciousness in the formation of reality, show that a "delayed reduction" of the universe by conscience might be possible; that is, that it is possible to change (or rather determine by an act of observation performed a posteriori by a "conscious being") the past after it has apparently happened!

On this subject, let us see what the well-known Cambridge astronomer, Rees, writes:

> In the beginning there were only probabilities. The universe could only have begun to exist if someone had observed it. It is not important that observers popped up several billion years later. The universe exists because we are aware of its existence.

1.20. Classical logic, quantum logic and the "uniqueness" of experiments

With Aristotle, Logic was officially born, the science of demonstration capable of indicating exactly when and why an argument is consistent, true or false, well-constructed with respect to its premises. Since Aristotle's time, everyday problems have been tackled and solved using the logic of "yes or no", "true or false". This bivalent logic is essentially based on the principle of *non-contradiction* – a thing cannot *be* and *not be at* the same time – and on the principle of the excluded third – a third possibility between the true and the false is not admissible: *tertium non datur*. Because of its importance, the principle of *non-contradiction* is described by Aristotle as "the strongest of all principles"; its negation would make thought and language impossible, since every concept could allude to one thing and at the same time to something else, even the opposite.

Certain situations in which the *superposition principle is* at work (see *two-slit* tests, experiments with Mach-Zehnder interferometers, etc.) experience a violation of the principle of *non-contradiction*. The possibility that a particle can be in two different places at the same time, that it can travel different paths at the same time, represents the coexistence (albeit *of a certain type*) of antithetic states. In quantum theory, moreover, those groups of statements that Bohr-Heisenberg describe as "meaningless" and that essentially refer to *indeterminate* states (to the unobserved states of matter), lying somewhere between the values of truth and falsity, effectively deny universal validity to the principle of the *excluded third*.

Not wishing to exclude from common speculative language, the *significant statements*, the assertions concerning indeterminate states,[18] can be applied a rule that does not allow them to be considered true or false.

This is achieved by introducing a third truth value; it is achieved by having *a three-valued logic*. To ordinary logic, therefore, for the

---

18   The meaning of "indeterminate state" is totally different from the meaning of "unknown". The term "unknown" is normally also applied to two-valued statements; the truth value of a statement of "common" logic may be unknown but belong to the group of true or false statements.

case of quantum mechanics, a third value must be "associated"; a neutral truth-value; which in the final analysis will make it possible to include indeterminate states among the statements with meaning.

The significance for quantum theory of the undetermined truth value is made particularly evident in the following considerations. Imagine a generic physical state S in which a measurement of the quantity X (e.g. position) is made, but in doing so one is forced to give up the knowledge of what the result would have been if a measurement of the "conjugate quantity" V (velocity) had been made. It is pointless to make a measurement of V in the new physical state since the measurement of X has changed the situation. It is equally useless to construct another system with the same initial state S, and to make a measurement of V, since the result of that measurement is only determined with a certain probability. This repetition of the measurement may yield a different value than the one that could have been obtained in the first case. The probabilistic character of the predictions of quantum mechanics entails an absolutism for the individual case; it makes the individual event unique and unrepeatable.

A similar situation to the one just illustrated occurs in the case of one person, we will call her Anna, who states: "If I throw a die on the next throw, I will get three" and another person, we will call her Bruno, who states: "If I throw the die instead, I will get four". Anna throws the die and gets *two*. We then know that Anna's assertion was false. As for Bruno's assertion, however, we have no possibility of judgement, as Bruno could not physically throw the dice. One way of verifying Bruno's assertion would be to measure the initial position of the die, the state of his muscles, the density of the air, etc. We could then predict the result of Bruno's throw with any degree of accuracy; or, better still, since we could never do it, Laplace's *superman* could do it for us, which in principle can reproduce any situation.

For quantum mechanics, however, not even Laplace's *superman* can fully reproduce the outcome of Bruno's launch, as, pointed out earlier, quantum physics experiments are governed by the law of probability and therefore the result of any observation is in principle uncertain, "variable" and therefore not reproducible.

1.21. Non-localism between actions at a distance and non-separability

Non-localism (i.e. the possibility of "distant influences") is one of those special phenomena that characterise the dynamics of quantum mechanics. In *ordinary* reality, influences between distant systems never occur directly or in real time. A flu epidemic that originates in Asia, for instance, does not immediately spread to Europe. It takes some time for it to reach our continent; it takes weeks for infected individuals to spread the disease as they move from one place to another on earth. The example (of *localism*) just given, does not apply in the world of quantum physics, where *actions* between different places in space are the order of the day, where reality seems to enjoy proving itself bizarre. To get a clear idea of quantum *non-locality,* imagine you have two boxes in front of you, each containing a glove from the same pair. It is "obvious" that, even before you look inside the boxes, you are certain that they will contain gloves with a well-defined "direction": the box on the right, for example, may contain a "right" glove, the box on the left a "left" glove, or vice versa. Now if, instead of using normal gloves, we used "quantum gloves", we would realise that the "direction" of the gloves in the respective boxes would only be defined at the moment of looking inside one of them. The act of looking inside one of the two boxes gives reality to the glove cover; it gives, at a distance (i.e. *not locally*), a "direction" to the glove not being *observed* (at that moment).

According to the quantum paradigm, before *observation,* before a observer decides to look inside one of the boxes, the gloves *live in an* overlapping, *intertwined state*; a state that sees the gloves "mixed" into a single entity: a *left-right* glove. It will be recalled how in the orthodox interpretation of quantum theory, the objective characteristics of any micro-element or pair of micro-elements are only defined when an act of observation is performed, only the act of observation *resolves* the superposed state that characterises matter. Let us now consider the emission by an excited atom of a pair of related particles:[19] let us say two photons. The logic of the man in the street (classical logic) tells us that even if these two particles were taken one to the Moon and the other to Venus, any experiment conducted on the photon found on our satellite would have no effect on the photon found on Venus. This is what common sense tells us and what physicists Albert Einstein, Boris Podolsky and Nathan Rosen also said (in 1935) when, with the intention of demonstrating the incompleteness of quantum theory, they wrote a series of critical articles, including a very famous one entitled: "Can quantum-mechanical description of physical reality be considered complete?" (Einstein, Podolsky and Rosen, 1935). In particular, Einstein, concerning the non-localism inherent in quantum mechanics, insite in the formalism of this theory, wrote:

> An essential aspect of the things of physics is that at a certain moment they can assert their independent existence from each other, provided they are located in different parts of space. If one does not make this kind of assumption about the independent existence of objects that are far away from each other in space, physical thinking in the familiar sense becomes impossible [...] The following idea characterises the relative in-

---

19   Related particles are micro-entities (photons, protons, etc.) that have a common origin. They can be somewhat assimilated to the homozygotic twins of animal species. The disintegration of a sub-atomic particle such as a π meson or the transition from the I to the II energy level of a calcium atom produces pairs of correlated photons. One of the main characteristics of correlated particles is that they exhibit, depending on the experimental situation, "symmetrical properties": e.g. same "oscillation" (polarisation) on a "plane".

dependence of distant objects in space (A and B); an external influence on A has no direct influence on B. If this axiom were to be abolished, the formulation of laws that can be empirically controlled in the accepted sense would become impossible.

In the passage just quoted, Einstein is categorical in reiterating the absolute independence of distant objects in space, on pain of chaos in physics. The exponents of the orthodox school of quantum physics are diametrically opposed to this, and in the words of their most authoritative exponent, N. Bohr, they condemn: "Even if two [related] photons were on two different galaxies, they would still remain a single entity [...] and the action performed on one of them would also affect the other". Although using terminology that differs slightly from that of Einstein and his companions (Bohr, in fact, speaks of "oneness of matter", of "inseparability of matter", not of *non-localism*), the quantum scientists emphasise that the fate of any related particle must always be common within the spatio-temporal evolution of the pair. The opposition between the exponents of the Copenhagen school on the one hand, and those who recognised themselves in Einstein's positions on the other, came to an end in 1982 when Alain Aspect (and co-workers) of the University of Paris conducted a series of highly advanced experiments[20] which

---

20   At this point, historical objectivity demands that mention be made of the work of the great Irish physicist John Bell. In 1964, Bell formulated a theorem (Bell, 1964) which, if applied to certain experimental contexts, could provide the theoretical tools to settle the dispute between Einstein, Podolsky, Rosen and quantum physicists. Bell's theorem – with its inequality – succeeded, with an elegant mathematical construction, in transforming the dispute between localists and non-localists, which had seemed to most to be nothing more than a philosophical dispute, into something that could be subjected to specialised verification. The formulation of Bell's theorem can be summarised by saying that any local theory, which assumes that certain pairs of correlated particles separated and sent to distant detectors have properties defined even before being

unequivocally demonstrated the rightness of the positions held by quantum theorists.

---

tested, cannot reproduce the probabilistic predictions of quantum mechanics. In particular, Bell, with the inequality elaborated within his theorem, defines the terms of the incompatibility between the theses that predict that pairs of correlated particles always and in any case have objectively defined and local properties, as opposed to the assumptions of quantum mechanics, which predicts that the reality of the microworld must only be defined at the moment a measurement is made, and that in fact this measurement must have non-local effects. According to some authoritative scholars, the implications at the level of the nature of reality that emerge from Bell's theorem transcend the dispute between Einstein and quantum physicists, presenting us with a non-local world "a priori". This is not the place to go into the technical-formal details of Bell's work, here we can however – by quoting the opinions of a number of scholars – provide a summary account of an eminently non-local reading of the great Irish physicist's work. The first of the scholars whose opinion is reported is James Cushing, who writes:

> *Bell never developed any local, deterministic theory. But, without ever going into dynamic details, he has shown that, in principle, no such theory can exist [...]. Bell's inequality [with its violation] is in no way dependent on quantum mechanics. It rejects a whole class of (essentially) classical theories without even mentioning quantum mechanics. It so happens that the experimental results not only exclude the entire class of local, deterministic theories, but also confirm the predictions of quantum mechanics. (Cushing, 1994)*

Complementing Cushing's arguments are the physicist David Lindley (1997) and the physicist and philosopher David Albert (1993): "Even if we do not like Quantum Mechanics, even if we think that some other theory might eventually supplant it, we cannot go back to the old view of reality. It simply does not work: that is the real importance, that is the real message of Bell's theorem", "What Bell's theorem has given us is the demonstration that there is an inescapable authentic non-locality in the unfolding of the processes of nature, however one tries to describe it. This non-locality is, first and foremost, a characteristic of quantum mechanics itself, but according to Bell's theorem, it is also necessarily a characteristic of every possible mode of calculation."

Before and after 1927.
The pictures show Bohr and Einstein in Brussels
during the Solvay conferences of 1925 and 1930.

Physicist David Lindley (1997) writes on this subject: "In 1982 the solution finally came and was accepted as definitive by the world of physicists. Nature did not obey the laws of Local Realism, quantum mechanics celebrated its triumph, and our understanding of the reality of the natural world became more complex".

The experiments conducted across the Atlantic by Aspect involved photons of a related pair being separated and launched towards distant detectors. The detectors in turn were to register the photons after a "polariser" was randomly inserted along the trajectory of one of them, conditioning its direction. The result of Aspect's tests showed that when one of the photons deflected as a result of interaction with the "polariser" placed in its path, the other one instantly deflected as well, even though it was spatially separated (thirteen metres apart to be exact, an enormous space for subnuclear-sized particles).

The extraordinary fact, already foreseen and feared by Einstein as early as 1935 in his experiments at the University of Paris, turned out to be not so much the confirmation of *non-localism*, of "actions at a distance", as the evidence that these *actions* took place "in real time", as if there were an instantaneous transmission of information between the related particles: violating the insuperability of the speed of light.

Non-localism, or *non-separability* (or *Entanglement* as the Anglo-Saxons say), presents us with a space-time at the antipodes of classical conceptions, compared to what we are used to experiencing every day.

1.22. Alain Aspect's experiments

In 1982, Alain Aspect, with the collaboration of two researchers, J. Dalibard and G. Roger, from the Institute of Optics at the University of Paris, took up the challenge for a rigorous verification of the "non-localist" hypotheses of quantum theory. He produced a series of highly sophisticated devices in the field of optics-physics, which made it possible to resolve the dispute that had been pitting physicists who recognised themselves in the "classical" positions (Einstein, etc.) against the quantum physicists of the *Copenhagen school for* almost half a century. In the figure below, we see a schematisation of the equipment used by Aspect and collaborators in their experiments (here, due to the complexity of the devices employed, we shall only analyse a simplified version of Aspect's experiments).

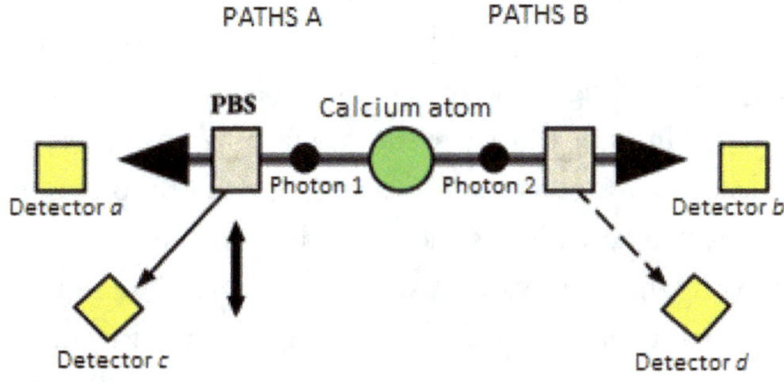

At the centre we have an excited calcium atom which produces pairs of correlated photons moving along opposite paths. A polarising optical beam splitter (PBS) is randomly inserted along one of these paths (Path A in the case shown in the figure), which, once a photon of a pair with a given polarisation interacts with it, can either be deflected or continue undisturbed on its way. A polarising optical beam splitter is also inserted along Path B, but in this case the PBS is not moved. At the ends of each potential pathway for each photon, a photon detector is placed.

Now, the extraordinary thing that Aspect has verified with its equipment is that at the moment when the PBS was inserted along Path A and a deflection of photon 1 (which at that moment had a certain polarisation: e.g. horizontal) was produced towards detector c, symmetrically photon 2 (i.e. the photon of Path B) also instantly deflected towards detector d. *In* practice, the act of inserting the PBS, with the consequent deflection of photon 1, caused photon 2 to be instantly and remotely deflected.

Aspect also verified that the action at a distance could also occur with "intermediate" photon polarisations (i.e. not only horizontal, but also at 22.5°, 45°, etc.), with the result that the deviation of photon 1 towards detector c induced (with a certain quantum probability) photon 2 to continue straight towards detector b.

This may sound strange, but it is what actually happens when experiments are performed on pairs of correlated particles. These eccentricities of nature (valid not only for photons but also for massive particles), quantum physicists stigmatise, are only so if one reasons according to "classical logic". In a scenario where one imagines that any related system (or particles that have interacted in the past) can enjoy the prerogative of not being affected by spatial distance, everything is simplified, "normal". By abandoning the idea that related particles located in distant places represent distinct entities, most of the conceptual (and factual) obstacles that prevent non-local *communication* or "action" also disappear.

Referring to the *uniqueness of matter* that results from the non-localist view of quantum theory, Nobel Prize winner for Physics Brian Josephson put it this way: "The universe is not a collection of objects, but an inseparable network of vibrating energy patterns in which no component has reality independent of the whole: including the observer in the whole".

1.23. Quantum physics yesterday and today

In closing this exposition of the foundations and implications of quantum theory, it is considered useful to repropose a few pages from a famous book written by Louis de Broglie (1969). Through these pages, the reader will have the opportunity to *relive* the intellectual climate (often made up of bitter contrasts) that characterised the emergence of quantum mechanics on the scientific scene from 1926 onwards. The lines written by Louis de Broglie will also make it possible to grasp, in the position of the physicists of the first half of the last century, the same approach of today's scientists towards the most "surprising" implications and conclusions of quantum mechanics.

At the end of October 1927, the fifth Council of Solvay Physics devoted to wave mechanics [here the term wave mechanics refers to quantum mechanics] and its interpretation. I [Louis de Broglie] gave an exposition of my attempt [to refute the probabilistic interpretation of quantum mechanics], but for various reasons I did so in a somewhat monotonous form and insisting mainly on the hydrodynamic image.

> My report was not at all appreciated: united around Bohr and Born, the very active group of young theorists, which included Pauli, Heisenberg and Dirac, was completely won over by the purely probabilistic interpretation of which they were the authors.

However, some voices were raised to curb these new ideas, Lorentz affirmed his conviction that the determinism of phenomena and their interpretation by means of precise images within the framework of space and time should be preserved; but his very notable intervention did not bring any constructive element [...].

What would Einstein have said in this debate from which the solution to the formidable problem that had worried him so much after his brilliant insight into the quanta of light was to emerge?

To my disappointment he said almost nothing. Only once did he speak for a few minutes: rejecting the probabilistic interpretation he did so in very simple terms. An objection that I believe nevertheless retained its considerable weight. Then he relapsed into muteness. In private conversations he encouraged me in my endeavours but did not comment on the theory of the double solution, which he did not seem to have studied very closely.

He claimed that quantum physics was on a bad path and seemed discouraged by this development. One day he said: These quantum physics problems become too complex. I cannot study such difficult questions any longer: I am too old! A very strange sentence in the mouth of this illustrious scientist who was then only forty-eight years old and whose bold thinking was certainly not one to be easily discouraged by the difficulty of problems!

I returned from the Solvay Council very disconcerted by the reception my ideas had received there. I did not see a way to overcome the obstacles they encountered and the objections that had been made to me. I had the impression that the current, which led the near unanimity of qualified theorists to adopt the probabilistic interpretation, was irresistible. I therefore aligned myself with this interpretation and took it as the basis of my teaching and research. The only attempt that had been made (the only one that could possibly have been made) to solve the problem of waves and corpuscles, in the sense that Einstein had hoped for, seemed to have decisively failed.

However, Einstein did not give up. As he emigrated to the United States, he did not cease to address in all his writings lively criticism of the purely probabilistic interpretation of quantum mechanics. There were lively clashes between Bohr and him, particularly in 1935 [...].

In this dispute, Einstein found himself almost isolated and had on his side only Schrödinger, the author of numerous and very fine objections against the probabilistic interpretation. But Einstein's attitude remained purely negative: he rejected the solution to the wave and corpuscle conundrum that had prevailed, but did not propose any other, which obviously weakened his position. He then devoted himself to his research into unitary theories which, prolonging the effort made by the development of general relativity, sought to conglobate the gravitational field, the electromagnetic field and possibly other fields into a single image. But even if these interesting unitary attempts were to continue to develop, they could in no way lead, at least in their present form, to an exact representation of physical reality, because they did not contain quanta. Einstein knew this better than anyone, he who had discovered the quanta of light! He was well aware that he could be reproached for having abandoned all constructive work in the quantum domain: and in his last letter to me, dated 15 February 1954, he jokingly said: I must resemble an ostrich that continually hides its head in the relativistic sand in order not to have to look these villainous quanta in the face.

Meanwhile, the controversy was slowly dying down. Most physicists accepted the probabilistic interpretation without question, some because they found it genuinely satisfying, others more pragmatic, because the formalism of quantum mechanics seemed to them to provide all the tools they needed for their predictions and because they no longer cared in any way what reality might be hiding behind the veil of equations.

It is in this last sentence that one can recognise the current position of the scientific community with regard to quantum mechanics, with regard to the more "surprising" implications of the Copenhagen paradigm.

For physicists, quantum theory is very useful for describing phenomena, making predictions and formulating new theories, but worrying about what actually lies behind *the veil of equations*, what lies behind the model of *reality* that emerges from the quantum-mechanical paradigm, is beyond the primary interest of research.

CHAPTER II

THE UNIVERSE FROM NOTHING

Since everything is emptiness, everything can be.
Nagarjuna

*Science must provide a mechanism that makes
account of the origin of the universe...*
John A. Wheeler

2.1. Nothingness, mathematical relations and the laws of physics

Why does something exist instead of nothing? It is an old, celebrated question that enjoys a bad reputation for its pitfalls.

Part of the problem lies in the fact that it is not clear what a satisfactory answer. An explanation, to be valid, must start from some point; the question, on the other hand, seems to demand that everything be explained immediately, without reference to what already exists. It is precisely for this reason that many scientists and philosophers interested in science literally refuse to answer questions about the "nature" of nothingness.

Eastern philosophical traditions (particularly Indian ones) regarded nothingness as a state from which anything could have come and to which it could return: indeed, these transitions could occur many times without beginning and without end.

Western philosophical traditions, on the other hand, sought to eschew nothingness. The Greek philosophers played a decisive role in conditioning Western thinking about nothingness. Both Plato and Aristotle, for entirely different reasons, had denied the existence of nothingness. Parmenides was even more radical with regard to the concept of nothingness, going so far as to affirm that one can only speak of what is: what is not (nothingness) cannot be thought and what cannot be thought cannot be.

Modern physics identifies nothingness as the absence of space-time and matter-energy (or ultimately, of fields). While imagining the absence of matter-energy and time may be within our mind's grasp, forming a mental image of an entity – nothingness – without a "place" in which to "circumscribe" it is improbable. Especially if we compare the absence of space that characterises nothingness with the space occupied by our universe. Where can nothingness be located outside our universe? The answer, perhaps, has to be confined to the realm of metaphysics and probably thinking about nothingness is beyond the reach of the human mind.

If nothingness cannot be thought of, however, that does not mean that there cannot be an entity that can be associated with nothingness: the number zero for example. The introduction of the zero symbol by the Indians owes much (as already mentioned above) to their ready acceptance of a multiplicity of concepts of nothingness and emptiness. Indian culture already possessed a rich network of concepts of nothingness that were in very common use. The creation of a numeral to denote a null quantity or an empty space in an accountant's ledger was a step that could be taken without having to redefine significant parts of a larger philosophy of the world. The use of the point symbol of zero in meditation exercises showed how a state of non-being was for Buddhists something to be actively sought in order to achieve Nirvana: unity with the cosmos. The hierarchy of Indian concepts of nothingness constitutes a coherent whole that organically includes the mathematician's zero symbol.

The Indian zero symbol reached Europe via Arabic culture, mainly through Spain. The Arabs were in close trade relations with India and thus came into contact with the efficient calculation methods developed there. Gradually, they incorporated the Indian zero into the notations of their complex mathematical-philosophical system.

It has been said that nothingness can be identified with zero, but among the many "uses" or applications of zero there is one that is extraordinary. Zero, and therefore nothingness, has the possibility of generating the number one and from the number one all numbers up to infinity. How is it possible, how can nothingness generate something and even more so generate infinity? To answer this question, one simply starts from zero, nothing else, and does perhaps the only one of the few natural operations that can be performed: raising zero to the zero power. Well, zero raised to the power zero does not equal zero, but one. This is the extraordinary power of zero that generates the number one from itself, that is, from nothingness.

At this point, there is not even a need to resort to obscure mathematics or philosophies such as, for example, that of the Chinese Lao Tzu, who maintained that the Tao generates the one, the one generates the two, the two generates the three and the three generates all things, because there is a theorem reformulated in the 20th century by the great mathematician John von Neumann, which shows that zero can generate an infinity of numbers.

To demonstrate this we will start with the most authentic symbol of zero, which in set theory is not zero itself, but an empty set. What does this empty set correspond to? To zero, by definition. Such an empty set is called an "original set". In mathematical language, the cardinality (i.e. the total) of the original empty set is said to be zero. Let us go a step further and see the process that, starting from zero, allows us to "create" all numbers. This will also give the most plastic picture possible of how "something" can come into being from nothing.

What do we have then? The empty set and zero, which is equal to the empty set. What is it possible to do now? Banally place the zero inside the empty set. At this point, an entirely new situation is created, because our set is no longer empty: it contains the zero, i.e. an element. The leap forward is remarkable, because from this moment on, the cardinality of the set is no longer zero, but one. A great result has thus been achieved: generating one from zero. Let us now take the number one and place it close to zero, within the set that was originally empty. This means, however, that there are now two elements in the set. The cardinality is therefore two. Thus the number two has been produced. And for the next number? The procedure is the same, so put zero, one and two in the original set and you get three. One can then proceed in the same way for the subsequent units, up to infinity.

$$\text{(Empty set)} = 0 \qquad (0) = 1 \qquad (0\ 1) = 2 \ \ldots\ \infty$$

Starting from zero, all natural integers were thus created. And what applies to the integers also applies – by introducing a few additional elementary laws – to all other families of numbers.

We have thus seen how from zero, that is, from nothing, something can be born. Let us now move on to analyse the possibility that something "tangible" could be born from nothing, something that could be the seed for the emergence of a universe like ours. To do this, we need to start with the laws of physics before we start with matter, because it is these that could allow nothingness to generate something.

2.1.1. Everything that is "possible" happens

We have already seen how all numbers can come from nothing, but like the laws of physics, mathematical relations (which, for example, allow zero to the power zero to generate unity) are also laws. And so the question that arises is where do the laws of physics come from, where do the laws that "govern" mathematical relations come from? The answer, perhaps, is less complicated than one might think. Let us try, for example, to change the name of zero and call it orez.

Orez raised to the power of orez will always give something that we might not call one but eno. So "nothing" raised to the power of "nothing" always gives "something" (specifically eno).

The key thing, then, is not so much what to call the entities that come into play in mathematics, but the relationship that binds them. In this case, however, the relationships, the laws, may have a kind of life of their own regardless of the "place" in which they take shape. Think of the Platonic idea that laws are out there and transcend physical reality: by virtue of being immaterial, laws can transcend reality.

Let us go one step further. If it is possible to assume that there are an infinity of laws, of mathematical relations, then all "possibilities" can occur, and this regardless of the entity where they are realised. Certain mathematical relations – such as the zero raised to the power zero which gives one – are possible because there is a non-zero possibility of this being realised. Mathematical relations, therefore, transcend reality and, like the infinity of numbers that arise from nothing, can have infinite ways of manifesting themselves. The laws of physics, like mathematical relations, transcend space, time, energy and can "emerge" from the infinity of possibilities potentially granted to realise themselves.

So, to summarise what has been said so far, we can state that the laws of physics, the mathematical relations, simply exist in potency "suspended above nothingness" and have infinite possibilities of expressing themselves, of passing from potency to act. There is nothing to prevent, therefore, if we start from the idea that there are potentially infinite possibilities that a given "order" can randomly or spontaneously arise out of nothing, that something like a law will manifest itself.

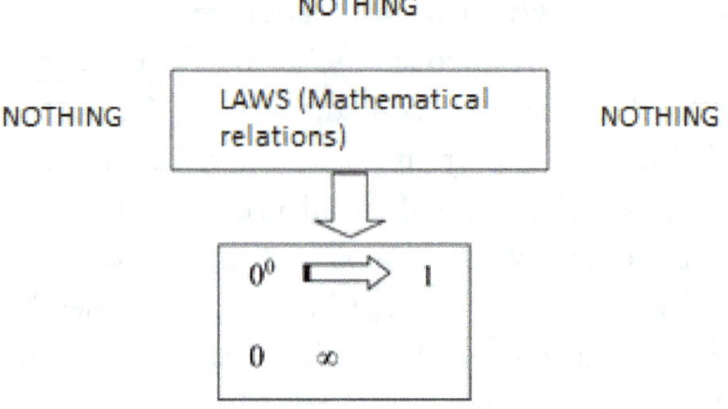

In the next section, we will see how, from the infinity of possible laws of physics (in this case, possible stands for plausible), the laws of quantum mechanics that allow a field to arise from nothing can be realised.

The attentive reader will have noticed that when referring to the universe, the terms "a universe like ours" are used. If laws are a potentially infinite number, our universe can only be one among many potentially possible universes. A universe like ours may have originated according to specific laws, according to certain constants, other universes may have originated according to opposite laws, and all of this by chance. In our universe, the fundamental constants of physics are these because the laws that shaped the early stages of the universe took the "shape" we know by chance. Just as there are many planets at many distances from our sun, and we live on a planet whose orbit produces conditions favourable to life, there are many universes whose fundamental constants have many different values; values that can generate life forms based on the most diverse chemical elements, e.g. silicon. There are therefore other universes, in fact there are potentially infinite universes, and we find ourselves in a particular universe because it is among those that support our form of life, because it is among those in which the laws of physics contemplate quantum mechanics. The point is that the property of existing does not make a particular universe special because all possible universes can exist in the *multiverse.*

2.2. Quantum mechanics and the emergence of a "field" out of nothing

This section will show how, starting from nothingness and the laws of quantum mechanics, a field can be born. A field that can be the seed for the birth of a universe such as the one we know today. Before getting to the heart of the matter, however, it is considered useful to make a few considerations regarding the mechanisms of the emergence of a universe from "nothing".

It must first be said, as already pointed out in the preface to this volume, that there are a number of books and articles in which the birth of the universe is hypothesised to be due to fluctuations in the "quantum vacuum". In these works, however, the mechanisms that account for the existence of the laws are not sufficiently explored in depth; regulatory laws are almost taken for granted without specifying where they come from. Asking where the laws come from and how they originate, on the other hand, is fundamental; only with the laws (particularly the laws of physics) does it become possible to hypothesise the genesis of the universe. Outstanding astrophysicists such as Tryon, Hawking, Hartle, and Vilenkin have formulated very interesting theories on the genesis of the cosmos; unfortunately, nothing emerges from their arguments about the birth of the laws that, among other things, they make extensive use of. In the present book, as we have seen, the genesis of laws comes first, preceding all other speculation. Moreover, in almost all the works on the origin of the universe, the subtle distinction between emptiness and nothingness is glossed over; that is, it is not made clear that emptiness is something other than nothingness.

Nothingness, from which any hypothesis of the genesis of a universe should start, is in fact the total absence of space, time and fields; the quantum vacuum, on the other hand, is a teeming of virtual fields and particles. To hypothesise, therefore, a universe arising from the quantum vacuum is to assume that something sees the light in a space where at least one field is already present. And here the question that arises is: by what mechanism does the field contained in the quantum vacuum come into being?

The Uncertainty Principle and the QED (as seen in Chapter One) exploit the fluctuations due to the energy-time uncertainty, allowing "something" to be born from the vacuum, but at least one field must be present in this vacuum.

How, then, can a universe that arises from the quantum void, i.e. a "full entity", be reconciled with a universe that must arise from something that does not pre-exist?

In the light of what has been said so far, it is understandable how much confusion reigns, even among the most influential experts on the subject. In this connection, let us see what the well-known astrophysicist John Gribbin writes under the heading "Quantum Cosmology" in his Encyclopaedic Dictionary of Quantum Physics:

> The universe as we know it must have been born with an age equal to the Planck time, the smallest unit of time that can exist, and must have had the Planck density (about 10^{94} grams per cubic centimetre). This kind of germ of the universe could have manifested itself from absolute nothingness as a quantum fluctuation, just as a pair of virtual particles can be produced from nothingness on the basis of energy borrowed from quantum uncertainty.

Now, anyone who has carefully read what has been written from Chapter One up to this point, may object that the nothingness Gribbin speaks of (nothingness in which quantum uncertainty would act) is nothing other than the quantum vacuum, but this vacuum, we cannot repeat often enough, is something else compared to nothingness. The quantum vacuum is full of fields, nothingness is the total absence of space, time and above all fields.

The aspect concerning the importance of defining the exact conditions under which our universe originates is overcome, in this volume (we shall see in a moment), by resorting to the possibilities that quantum formalism offers concerning a phenomenon we have already encountered, namely the "superposition of states".

Still on the subject of the emergence of a universe with the mechanism of vacuum fluctuation, it is worth mentioning the conclusions reached by several Anglo-Saxon scholars to make a universe that sees light from the vacuum stable with a mechanism similar to the emergence of virtual particles. These scholars invoke the retroactive intervention of an observer's conscience, as outlined in Sections 1.18 and 1.19 of Chapter One, to make permanent and stable a universe that has arisen from the vacuum and should only live for a few moments.

Another possibility for the birth of the universe from nothingness identified by some scholars involves the "Many Worlds" interpretation (see Section 1.14 of Chapter One), where a new "World' arises – apparently from nothingness – from the "splitting" of our universe following an "act of measurement". In this case, however, it is not specified where the "energy" that gives rise to a new universe comes from, and then before the "splitting", i.e. before a new "World" appears, there is already something, there is already our universe; hence, it is trivially impossible to conjecture the birth of a universe from nothing. Having closed the parenthesis concerning the main hypotheses predicting the birth of the universe from the vacuum, we return to the theses advanced in the present book.

In quantum mechanics, as we saw in Chapter One, there are very particular situations in which superposition of states comes into play. The superposition of states can be applied in numerous quantum fields where one has to deal with dual aspects of matter: one of these, for example, is the wave and corpuscular aspect of a photon, an electron, etc. In this case, the mathematical formalism applied is very simple; in fact, the wave function (Ψ) of the particle is always equal to the sum of the wave functions of the two sub-states (wave-corpuscle) under consideration at the time. Now, if we were to apply the typical formalism of superposition of states to a field in which the dual states at play are the "vacuum" and "non-vacuum" sub-states, we could obtain some very interesting results. Indeed, from the formula: Ψ (final) = Ψ (void) + Ψ (non-void), which represents the superposition of "void" and "non-void" (in this case it should be pointed out that void should not be understood as a quantum vacuum, but as "pure nothingness"), we can imagine that the state of "void" prevails with a certain probability, but, equally, *we cannot exclude* that the state of "non-void" may also prevail. Here, out of nothingness it is possible to obtain something tangible, among other possibilities, a field; a real "primordial energy field" that can generate the energy and inflaton[21]

---

21   The term inflaton, which denotes both the field and the associated particle, is derived from "inflation", i.e. the rapid expansion of space in the very first moments of the universe's existence. Cosmic inflation is based on the idea that the universe initially passed through the "false vacuum" state of a quantum field called the inflaton. This field that permeated the entire universe was the only one that existed at the time; common fields and their associated particles would appear much later. Like the Higgs field, the inflaton had an energy profile with a local maximum when its configuration was in a state of "false vacuum", a state with no particles associated with the field. When gravity is introduced in this context, it is observed that the inflaton, in its false vacuum state, generates a strong gravitational repulsion, resulting in the abnormal expansion of space through inflation. It is therefore ultimately assumed that in the first moments of the universe's life, space was filled with a field, the inflaton, that was in a state of "false vacuum" which, thanks to quantum fluctuations of its own, emerged from its primitive condition to converge into a new state that led to a modulation of inflation to a standard expansion rate allowing the creation of common fields, plasmas and particles. Standard cosmology thus originated.

(or a Higgs field[22]) that leads to the birth (identified with the Big Bang) of a universe like ours.

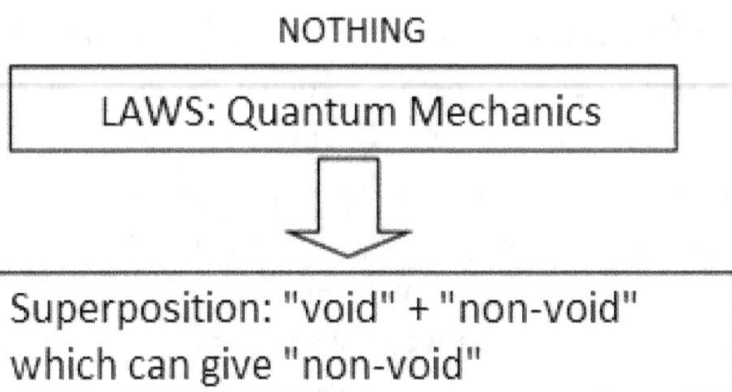

Now one objection that could be raised is that in order for a superposed state to obtain a well-defined state (i.e. for there to be a reduction of the wave function), an act of observation, a measurement, is required. This is true in a certain sense; however, the fact that one cannot cause the reduction of the wave function without an observation does not mean that there cannot be such a reduction. On the contrary, anything that is not "in power forbidden" is in fact possible. Hence, the mechanism that allows the "non-void" to emerge from the superposition of states can also be achieved without a measurement. As demonstrated, the state of emptiness can be thought of as a superposition of states that are not void, nothing prohibits, therefore,

---

22   In 1964, at the dawn of the Standard Model of particle physics, several scientists came up with a theoretical model with which one could explain, in a coherent way, why some particles had mass and others did not. This was the so-called Higgs mechanism (named after the name of British physicist Peter Higgs), which predicts the existence of a field, the Higgs field, and its associated particle, the Higgs particle. The existence of said particle (and indirectly Higgs field), was experimentally demonstrated in the LHC accelerator of the CERN in Geneva in the year 2012. The experimental confirmation of the Higgs field was a decisive event for physics, in fact the presence of the Higgs field-particle is able to explain the mass of particles.

that (albeit with a very low frequency) the "non-void" can "prevail" ("spontaneously") over the void regardless of an act of measurement.

An idea of how a spontaneous reduction of the superposition of two different states can occur can be had by thinking back to the solution that Roger Penrose in Section 1.14 of Chapter One gives to the Measurement Problem. The British scholar hypothesises that spontaneous (what he calls "orchestrated") collapses of two different superposed states of the space-time metric can occur in nature so that a well-defined state can spontaneously "emerge" from a superposition. Also in Section 1.14 we have another example of how a spontaneous reduction of the wave function can occur; this is the theory developed by GianCarlo Ghirardi, Alberto Rimini and Tullio Weber. The three Italian physicists believe that with the passage of time, the wave function of a microscopic entity tends to "expand"; during this expansion, there is the possibility that spontaneous reduction of the two different superposed states that characterise the microscopic entity may occur.

2.3. The "birth" of time and space

Two problems concerning the birth of the universe now remain to be addressed and solved. The first problem concerns the moment in which time "begins" to flow, the second problem concerns the moment in which space "takes shape".

The problems concerning the emergence of time and space, as well as their definition, have caused rivers of ink to flow over the past centuries (we have discussed them at length in Part I of this book). Theologians, philosophers and scientists have formulated the most disparate hypotheses. When we think of space and time, however, we are accustomed to take them almost for granted; this is because we actually analyse these two concepts with a gaze bordering on the superficial. Space and time are there (they are the stage on which the world plays out its endless comedy) as something permanent, unchanging, that can always be counted on. When Einstein challenged the most common and entrenched beliefs about space and time (by introducing the concept of space-time) scientists could no longer avoid a radical overhaul of traditional models.

The Greek thinkers of ancient times had not conceived the notion of the of a development in the progressive direction of time; rather, it was observed that many natural phenomena had a cyclical organisation. Day was followed by night, and night was succeeded by day again. Seasons also succeeded one another in an annual cycle that was repeated again and again. As a result of these phenomena, time was not perceived as an omnipresent unidirectional flow – like a "river of time" – but was rather described as a cyclic alternation of opposites. For example, the mathematician and philosopher Anaximander of Miletus around 580 B.C. indicated the primary basis of existence in the infinite (or "indefinite", àpeiron) movement generating opposites: hot and cold, dry and wet; then everything had to return to its original state.

As an alternative to Greek thought, time was mostly considered as something flowing continuously, comparable to a river, flowing into the infinite past and the infinite future. The main characteristic of time, however, is that it is uniform and universal, devoid of any mutation, even slight: a kind of all-encompassing container of nature and its activities. Newton himself was very clear about the universal and absolute nature of time: "Absolute, real, mathematical time [...] always flows in the same way, without relation to anything external".

Even space has been considered for centuries as something of fixed immutable. Newton again stated: "Absolute space, for its very nature, devoid of relation to anything external, remains always similar to itself and immobile". James C. Maxwell, centuries after Newton, writing on matter and motion, conceived of absolute space as "something that remains similar to itself and immutable". In fact, Maxwell argued: "to say that space moves would be like saying that a place moves. distant from itself".

In the modern era concerning time, Einstein, with his theory of Relativity, abandons the concept of absolute time by theorising about a time proper to each "observer". Finally, in the last forty years, the concept of time has undergone another profound revision. Today, there are scientists who, for example, have circumvented the problem of the instant in which the universe "comes to life", arguing that time can arise (without a precise beginning) from a coordinate of space.

In a scenario such as the one hypothesised in this book, on the other hand, the birth of space and time must be considered contemporaneous with the emergence of the field. Space and time would therefore see their birth from nothingness at the moment when, by the quantum mechanism illustrated above, the "non-void" ("spontaneously") prevails over the void and generates a field; that's all!

With these last considerations, *the circle* opened by the famous question *"why is there something instead of nothingness?"* is closed; for "something always *inevitably* arises from nothingness".

2.4. The universe born out of nothing at the test of facts

Let us now address one of the fundamental problems of any theory; the experimental proof. With the theses advanced in the book, an attempt has been made to demonstrate how our universe could have arisen from the most absolute nothingness, from metaphysical nothingness. An attempt has been made to show how emptiness and nothingness are two completely different "entities". An attempt was also made to show how superficial it is to use laws, particularly the laws of physics, without asking where these laws come from.

One question that arises at this point is: is there any possibility of proving experimentally (beyond the purely theoretical level, which in itself can already be considered a consistent point of arrival) that a universe came into being from nothing? Cosmology tells us that the inflationary era has erased all traces of what came before the abnormal expansion of space. Perhaps, however, a possible verification of a universe born from nothing is possible.

In physics, fundamental quantities such as electric charge, energy and rotational motion (angular momentum) must conserve themselves. This means that these quantities cannot be created out of nothing or disappear into nothingness. If one starts from something that already exists and possesses certain characteristics, these characteristics must conserve themselves; conversely, nothing can be created and subsequently conserved in nothing.

These arguments are entirely plausible until we begin to ask ourselves what specifically electric charge, energy and rotational motion are in the context of our universe.

In the present universe, for example, the electrical charges of atoms (atoms that ultimately make up everything in existence) are in equilibrium overall. In fact, the number of electrons and protons in the atoms are equal. So the obvious conclusion to be drawn from this is that the overall electrical charge of the universe is zero.

Let us now consider energy – which for Einsteinian Relativity is equivalent to mass-energy. Energy is the most intuitive example of how something is opposed to nothing. Well, even for energy, it can be said that in a closed universe such as ours, it must be globally equal to zero. The reason lies in the balancing act with gravitational energy. Basically, the positive contribution of the energy in the universe is cancelled out by the negative contribution of gravitational energy.

The fact that the total gravitational energy in our universe is negative derives from the equations of General Relativity. Consider, for example, a stone that is thrown into the air. In the fall (due to gravitational attraction) towards the surface of the Earth, the stone will gradually lose energy. So the energy will decrease as the rock approaches the Earth.

Now if we assume that the potential energy of the stone is equal to zero when it is at an infinite distance it is evident that energy will become increasingly negative as the stone approaches the Earth. In fact, the potential to perform work decreases with the approach of the rock to our planet. In the end the situation protagonist the stone thrown into the air can be applied, with all the care, to every "object" in our universe in expansion.

The cosmos as a whole can therefore be compared to an entity whose global energy is constantly counterbalanced by negative gravitational energy. It follows that the total energy of the universe is zero.

This conclusion is now accepted by the majority of cosmologists and although it seems paradoxical, it brings yet another piece in favour of the thesis that our universe came into being from nothing.

Finally, it remains to be ascertained whether our universe possesses a rotational motion or whether the global angular momentum is zero.

The data in the possession of the scientific community tell us that the cosmic background radiation has almost the same temperature in every region of the universe. If the universe rotated, the microwave radiation would be hotter if it came from the direction of the axis of rotation and colder if it came from directions orthogonal to it. The fact that the cosmic background radiation has almost the same temperature in every direction (with deviations of one part in a hundred thousand) shows that the angular momentum of the universe must be zero.

So what conclusions can we draw from the fact that the cosmos as a whole has zero electrical charge, zero energy and zero rotational motion? Certainly not that the cosmos definitely arose from nothing. It does, however, add an important tile towards the plausibility of a scenario in which the universe arises ("spontaneously") from nothing by the mechanism described in this book.

2.5 A possible future for our universe

As the book closes, a question almost arises: is a universe born out of nothing, a universe born through the mechanisms described so far (and which at this point we venture to call the Sifr Model: the term Sifr in Arabic means both emptiness and infinity), stable or could it suddenly disappear as it came into being?

The standard cosmological model predicts three types of geometric surfaces that can describe the state of the universe:

a. A surface such as the Earth's that is closed and which mathematicians call *positive curvature*. A spherical universe with a mass density that exceeds a certain critical value will expand with a radius of curvature that will reach a maximum and then begin to decrease, and in a phase of contraction symmetrical to that of expansion (a Big Crunch) will thus eventually reach a null value, ultimately making space disappear in a finite time future. This is well represented by a cycloid (a curve born from the rotation of a sphere without a spatial dimension). It has been hypothesized that once the universe has reached the lowest point of collapse it could reverse itself and be reborn through a new Big Bang in a continuous succession of Big Bangs and Big Crunches.

b. Another type of curvature, but this time negative (hyperbola), is represented by the surface of a saddle which has *negative curvature* and is open, that is, it does not close again like the surface of a sphere. A universe represented by a hyperbolic curvature is open and continues indefinitely with the passage of cosmic time.

c. Finally, what mathematicians describe by a flat surface depicted, for example, by the plane of a table, has *zero curvature*. Such a surface has no curvature and therefore does not occupy the third dimension. A universe with zero curvature, as well as a universe with negative curvature, will continue to expand eternally by gradually diluting its matter until it becomes a cold, dark place that is essentially dead.

In the figure reproduced below we see represented the three types of universes so far described. The first has positive curvature and is represented by a cycloid. It is a closed universe with limited life (if we exclude the possibility of its oscillating variant).

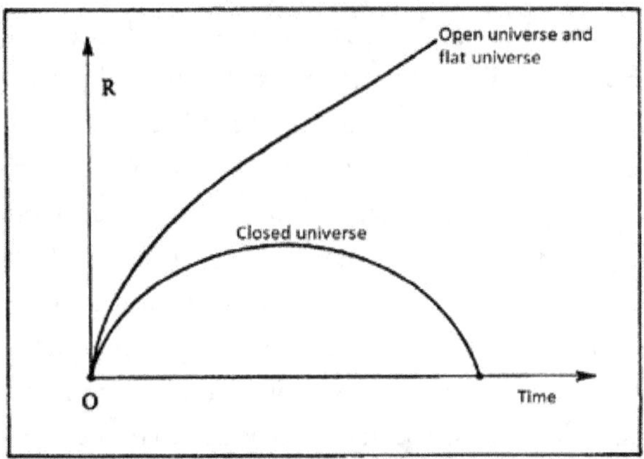

The universe Hyperbolic is an open universe with a lifetime of indefinite duration. A flat universe is a universe whose characteristic curve is similar (from its inception to the present) to the curve of a hyperbolic universe with a lifetime of indefinite duration.

Recent studies of the cosmic background radiation have shown that our universe is flat with an accelerated (current) rate of expansion.

In any case, in order to know with a good degree of approximation how our universe will actually evolve, it will be necessary to know "what it will do" (which is currently unknown) about the dark energy that accounts for about 70 percent of the "mass" of the universe.

How does it fit into the framework of a series of cosmic evolutions as those just frescoed, the Sifr Model, i.e., our model of the universe born out of nothing?

The fact that the universe has now come of age, that is, it has been in existence for as many as 13.72 billion years, might lead us to think that the cosmos will extend its life indefinitely.

But if we think carefully about how *our* universe in the hypothesis made in this volume came into being, it is not excluded that the universe at some point in its life may spontaneously "disappear" just as quickly as it appeared.

BIBLIOGRAPHICAL INDICATIONS

Albert, D. Z., *Quantum Mechanics and Experience*, Harvard University Press, 1994.

Bell, J., *On the Einstein-Podolsky-Rosen Paradox*, Physics, vol. 1, 1964.

Born, M., *Atomic Physics*, Hafner, 1957.

Cushing, J. T., *Quantum Mechanics*, Chicago University Press, Chicago, 1994.

de Broglie, L., *Nuove Prospettive in Microfisica*, Fratelli Fabbri Editori, Milan, 1969.

Einstein, A., Podolsky, B. and Rosen, N., *Can quantum-mechanical description of physical reality be considered complete?*, Physical Review, 47, 1935.

Heisenberg, W., *Physics and Philosophy*, il Saggiatore, Milan, 1961.

Lindley, D., *Where Does the Weirdness Go?*, Basic Books, Harper Collins, 1996.

Renninger, M., *Zeitschrift für Physik*, 158, 1960.

Schrödinger, E., *Die gegenwärtige Situation der Quantenmechanik*, Naturwiss, 23, 1935.

Selleri, F., *La Causalità Impossibile*, Jaca Book, Milan, 1987.

Wigner, E., *The scientist speculates*, W. Heinemann, London, 1961.

ANALYTICAL INDEX

INDEX

www.ingramcontent.com/pod-product-compliance
Lightning Source LLC
LaVergne TN
LVHW050736200726
843507LV00001B/22